Observations of distant supernovae have recently provided startling evidence that the expansion of the Universe may be accelerating, rather than decelerating. If this result is verified by future studies, it has profound implications for cosmology. The reliability of this finding and its implications for both the study of supernovae and cosmology are the subject of this exciting volume. Based on a conference at the University of Chicago, this timely volume presents articles by leading experts on the theory of Type Ia supernovae, observational astronomy, and cosmology. It examines the observational data, the nature of the likely progenitor binary systems, the outburst mechanisms of Type Ia supernovae events, and the cosmological implications.

This is a unique and wide-ranging review of one of the most dramatic and controversial results in astronomy in recent decades. It makes fascinating reading for all researchers and graduate students.

T0255232

CAMBRIDGE CONTEMPORARY ASTROPHYSICS

Type Ia Supernovae

CAMBRIDGE CONTEMPORARY ASTROPHYSICS

Series editors
José Franco, Steven M. Kahn, Andrew R. King and Barry F. Madore

Titles available in this series

Gravitational Dynamics,
edited by O. Lahav, E. Terlevich and R. J. Terlevich
(ISBN 0 521 56327 5)

High-sensitivity Radio Astronomy,
edited by N. Jackson and R. J. Davis (ISBN 0 521 57350 5)

Relativistic Astrophysics,
edited by B. J. T. Jones and D. Marković (ISBN 0 521 62113 5)

Advances in Stellar Evolution,
edited by R. T. Rood and A. Renzini (ISBN 0 521 59184 8)

Relativistic Gravitation and Gravitational Radiation,
edited by J.-A. Marck and J.-P. Lasota (ISBN 0 521 59065 5)

Instrumentation for Large Telescopes,
edited by J. M. Rodríguez Espinosa, A. Herrero and F. Sánchez
(ISBN 0 521 58291 1)

Stellar Astrophysics for the Local Group,
edited by A. Aparicio, A. Herrero and F. Sánchez
(ISBN 0 521 63255 2)

Nuclear and Particle Astrophysics,
edited by J. G. Hirsh and D. Page (ISBN 0 521 63010 X)

Theory of Black Hole Accretion Discs,
edited by M. A. Abramowicz, G. Björnsson and J. E. Pringle
(ISBN 0 521 62362 6)

Interstellar Turbulence,
edited by J. Franco and A. Carramiñana (ISBN 0 521 65131 X)

Globular Clusters,
edited by C. Martínez Roger, I. Pérez Fournón and F. Sánchez
(ISBN 0 521 77058 0)

The Formation of Galactic Bulges,
edited by C. M. Carollo, H. C. Ferguson and R. F. G. Wyse
(ISBN 0 521 66342)

Type Ia Supernovae,
edited by J. C. Niemeyer and J. W. Truran (ISBN 0 521 78036 5)

Type Ia Supernovae
Theory and Cosmology

Edited by
J. C. NIEMEYER
University of Chicago

J. W. TRURAN
University of Chicago

CAMBRIDGE UNIVERSITY PRESS
Cambridge, New York, Melbourne, Madrid, Cape Town, Singapore,
São Paulo, Delhi, Dubai, Tokyo

Cambridge University Press
The Edinburgh Building, Cambridge CB2 8RU, UK

Published in the United States of America by Cambridge University Press, New York

www.cambridge.org
Information on this title: www.cambridge.org/9780521126533

© Cambridge University Press 2000

First published 2000
This digitally printed version 2009

A catalogue record for this publication is available from the British Library

ISBN 978-0-521-78036-0 Hardback
ISBN 978-0-521-12653-3 Paperback

Contents

The Cosmological Observations

High Redshift Environments

Progenitor Models

Explosion and Light Curve Modeling

Cosmological Perspectives

Participants

David Arnett	University of Arizona, dave@bohr.physics.arizona.edu
Eddie Baron	University of Oklahoma, baron@mail.nhn.ou.edu
Ezra Boyd	University of Chicago, ecboyd@oddjob.uchicago.edu
David Branch	University of Oklahoma, branch@phyast.nhn.ou.edu
Martin A. Bucher	Cambridge University, M.A.Bucher@damtp.cam.ac.uk
Ramon Canal	University of Barcelona, ramon@am.ub.es
Gertrud Contardo	European Southern Observatory, gcontard@eso.org
Asantha R. Cooray	University of Chicago, asante@hyde.uchicago.edu
Alan Dressler	Carnegie Observatories, dressler@ociw.edu
Jonathan Dursi	University of Chicago, ljdursi@eugenia.asci.uchicago.edu
Alex Filippenko	University of California Berkeley, alex@wormhole.Berkeley.EDU
Josh Frieman	University of Chicago, frieman@fnal.gov
Peter Garnavich	Harvard University, peterg@mars.harvard.edu
Donald Goldsmith	Science Writer, dgsmith2@ix.netcom.com
Mario Hamuy	University of Arizona, mhamuy@as.arizona.edu
Kazuhito Hatano	University of Oklahoma, hatano@phyast.nhn.ou.edu
Wolfgang Hillebrandt	Max-Planck-Institut für Astrophysik, wfh@mpa-garching.mpg.de
Peter Höflich	University of Texas Austin, pah@alla.as.utexas.edu
Daniel Holz	University of Chicago, deholz@rainbow.uchicago.edu
Icko Iben	University of Illinois, icko@sirius.astro.uinc.edu
Koichi Iwamoto	University of Tokyo, iwamoto@lilas.astron.s.u-tokyo.ac.jp
David Jeffery	University of Nevada, jeffery@brewster.physics.unlv.edu
Saurabh Jha	Harvard University, saurabh@mondatta.harvard.edu
Ron Kantowski	University of Oklahoma, ski@phyast.nhn.ou.edu
Manoj Kaplinghat	Ohio State University
Alexei Khokhlov	Naval Research Laboratory, ajk@lcp.nrl.navy.mil
Robert Kirshner	Harvard University, kirshner@quincy.harvard.edu
Steven N. Koppes	University of Chicago, s-koppes@uchicago.edu
Don Lamb	University of Chicago, lamb@pion.uchicago.edu
Eric Lentz	University of Oklahoma, lentz@phyast.nhn.ou.edu
Martin Lisewski	MPI für Astrophysik, lisewski@mpa-garching.mpg.de
Mario Livio	STSCI, MLIVIO@stsci.edu
Rob Lopez	University of Chicago, lopez@aether.uchicago.edu
Tom Loredo	Cornell University, loredo@spacenet.tn.cornell.edu
Howie Marion	University of Chicago, hman@athena.uchicago.edu
Paolo Mazzali	Osservatorio Astronomico Trieste, mazzali@gandalf.oat.ts.astro.it
Peter Meikle	Imperial College, p.meikle@ic.ac.uk
Ana Paula Miceli	Fermi National Laboratory, anapaula@fnas12.fnal.gov
Coleman Miller	University of Chicago, miller@oddjob.uchicago.edu
Peter Milne	Naval Research Laboratory, pmilne@astro.phys.clemson.edu
Jens Niemeyer	University of Chicago, j-niemeyer@uchicago.edu
Ken'ichi Nomoto	University of Tokyo, nomoto@astron.s.u-tokyo.ac.jp
Peter Nugent	Lawrence Berkeley Laboratory, penugent@lbl.gov
Saul Perlmutter	Lawrence Berkeley Laboratory, saul@lbl.gov
Philip Pinto	University of Arizona, ppinto@as.arizona.edu
Jean M. Quashnock	University of Chicago, jmq@virgo.uchicago.edu
Paul Ricker	University of Virginia, pmr7u@virginia.edu
Adam Riess	University of California Berkeley, ariess@salmo.Berkeley.EDU

R.A. Rosenstein   lightcone@hotmail.com
Pilar Ruiz-Lapuente   University of Barcelona, pilar@mizar.am.ub.es
Nick Suntzeff   NOAO, nick@ctiow5.ctio.noao.edu
Ramon Toldra   Fermi Lab, toldra@fnal.gov
Jim Truran   University of Chicago, truran@nova.uchicago.edu
Mike Turner   University of Chicago, mturner@oddjob.uchicago.edu
Hideyuki Umeda   University of Tokyo, umeda@astron.s.u-tokyo.ac.jp
Nevin Weinberg   University of Chicago, nweinber@midway.uchicago.edu
Craig Wheeler   University of Texas Austin, wheel@alla.as.utexas.edu
Craig Wiegert   University of Chicago, c-wiegert@uchicago.edu
Stan Woosley   University of California Santa Cruz, woosley@ucolick.org
Scott E. Wunsch   CRF Sandia, wunsch@sandia.gov
Don York   University of Chicago, don@oddjob.uchicago.edu
Mike Zingale   University of Chicago, zingale@oddjob.uchicago.edu

Preface

Perhaps as we might expect, our perspective concerning the past and future of the Universe is ever changing. During the early years of this century, we first became aware of the fact that ours is an expanding Universe. The question which obviously follows is will it expand forever or rather ultimately collapse? Over the past several decades, we have searched in vain for some ubiquitous 'dark matter' (beyond that observed in stars and galaxies) that might provide sufficient gravity to halt the expansion. Most recently, the observed brightnesses of distant supernovae -together with the assumption that their behaviors precisely mimic those of their nearby counterparts - provide evidence that the expansion of the Universe is actually accelerating. It is the examination of both the reliability of this finding and its implications for supernova theory and cosmology that constitutes the subject of this workshop proceedings.

The tools of choice for these recent explorations of the rate at which the Universe is expanding are, specifically, supernova explosions of Type Ia. Occurrences of these and similar explosive stellar events have been recorded for millenia (as visiting stars) by careful observers of the night sky. Only in this century, however, has our improved knowledge of the distance scale of the Universe allowed us to distinguish between novae and supernovae on the basis of their intrinsic brightness. Two broad classes of supernovae are observed to occur in the Universe: Type I and Type II. Observationally, the critical distinguishing feature of Type I supernovae is the absence of hydrogen features in their spectra. Theory now focuses attention on models for Type I events involving either exploding white dwarfs or the explosions of massive stars which have, via wind driven mass loss or binary effects, shed virtually their entire hydrogen envelopes. We will be concerned specifically with supernovae of Type Ia, a subclass of the Type I's which are assumed to involve the explosions of white dwarfs in binary stellar systems. Such explosions are understood to represent the site of formation of most of the the nuclei in nature in the 'iron-peak' region, from titanium through zinc (Ti-V-Cr-Mn-Fe-Co-Ni-Cu-Zn).

The workshop itself, which took place at the University of Chicago in October 1998, brought together the leading experts on the theory of Type Ia supernovae to examine the question as to whether the use of the light curves for these cosmic explosions as a mechanism for determining distances was understandable and justifiable. The relevance of Type Ia supernovae to our determination of the distance scale of the Universe arises from the facts that (1) *they are the brightest stellar objects known* and (2) *they are good 'standard candles'* in that their peak brightness (luminosity), although not recognized to be precisely constant from event to event, appears to be correlated with the structure of the light curve. The observational basis for the observed relationship is nicely summarized in the article by Alexei Filippenko and Adam Riess.

The supernova events whose characteristics are being used to establish the distance scale are located at red shifts in the range $z \sim 0.5$ to 1. Questions concerning whether the Universe appears different at such distances are addressed in the articles by Alan Dressler and Don York.

Theoretical models for the origin of such Type Ia supernova events typically invoke rather involved epochs of mass transfer in close binary systems which can lead to the growth of white dwarfs to *the Chandrasekhar limit,* $\approx 1.4\ M_\odot$. Estimates of the time required for a specific binary system to yield a Type Ia supernova event range from 10^8 to 10^{10} years, which range allows Type Ia supernovae to occur - as they are indeed observed to occur - in both spiral and elliptical galaxies. The so called *'standard model' for a Type Ia event* involves the ignition of ^{12}C in the core of a white dwarf in a close binary system leading to: (1) the conversion of approximately 0.6 $M_\odot$ of the core to ^{56}Ni, the decay of which through ^{56}Co to ^{56}Fe on a timescale of approximately 113 days powers the supernova light curve; (2) the conversion of most of the remaining mass into nuclei from ^{16}O to ^{40}Ca; (3) the ejection of the entire core of ≈ 1.4 $M_\odot$; and (4) no condensed remnant (e.g. no neutron star remnant). We may note, however, that recent work considers as well models for Type Ias which involve *"sub-Chandrasekhar" mass* white dwarfs. Detailed discussions of Type Ia supernova progenitor models are presented in the papers by Mario Livio and by Pilar Ruiz-Lapuente and her collaborators. Questions concerning explosion models and their possible dependencies upon such environmental factors as the heavy element content of the gas from which the progenitor stars were formed at early times (higher red shift) are examined in the papers by Ken Nomoto and by Peter Höflich and his collaborators.

The observational data seem to indicate that supernovae Ia are systematically dimmer at increasing red shifts, implying the existence of a cosmological constant. Extant theoretical models cannot readily explain how SN Ia explosions might "mimic" a cosmological constant by making the events systematically dimmer at high redshifts. What are the implications for cosmology? These issues are addressed in the papers by Michael Turner and Josh Frieman. In a nice summary, Wolfgang Hillebrandt asks rather whether, in the context of supernova theory, $\Lambda = 0$ models can be salvaged.

During the eleven months since the time of the workshop, a number of new developments took place regarding possible systematic effects influencing the cosmological measurements. In particular, three recently published investigations would have fit well into the framework of this volume. We therefore briefly sketch their main ideas below.

Aguirre (1999abc) proposes the existence of a uniform cosmological dust component, ejected by bright, dusty, starburst galaxies at $z \sim 1$, and composed of large ($\geq 1\mu$m) dust grains that would give rise to grey extinction without appreciable reddening. Independent of the supernova observations, this idea is motivated by a suspected metal enrichment of the intergalactic medium by ejection of metals and dust from galaxies via winds and the preferential destruction of small dust grains. It can be tested by observations of supernovae at $z > 1$ which should appear systematically dimmer in the dust model while the effect of a cosmological constant is predicted to reverse sign at higher redshifts. A sufficient amount of Aguirre's dust could reconcile the observations with an open $\Lambda = 0$-universe but not with a standard $\Omega = 1$ CDM universe.

Drell, Loredo and Wasserman (1999; also reported by Tom Loredo at the workshop) present a statistical analysis of the supernova observations based on Bayesian methods that accounts for the possibility of simple phenomenological models for supernova evolution as a function of redshift. They conclude that only if evolution is assumed not to occur, the data unequivocally favors a non-zero value of Λ. If the possibility of evolution is taken into account, the authors find that if the geometry of the universe is left unspecified, the observations do not discriminate between a universe with $\Lambda = 0$ or $\Lambda \neq 0$. Furthermore, they argue that two indications hint at a possible systematic difference of nearby and distant supernovae. The first is the mutual inconsistency of two different brightness correction schemes at high redshift which seems to be absent in the low-z sample. A second hint, according to the authors, is the fact that the correction schemes reduce the dispersion of peak supernova brightnesses in the nearby sample while failing to do so at high redshift.

Finally, and perhaps most importantly, there are now first observational indications of evolutionary differences between low and high redshift SNe Ia. Riess et al. (1999) have compared the rise times of both samples and found a difference of 2.5 ± 0.4 days with a statistical confidence of 5.8σ. If these findings are confirmed, the possibility that the evolutionary process that gives rise to the rise time difference also affects the peak luminosities must be taken under serious consideration.

REFERENCES

AGUIRE, A. N. 1999a, *ApJ*,**512**, L19.

AGUIRE, A. N. 1999b, *ApJ*, in press (astro-ph/9904319).

AGUIRE, A. N. 1999c, *ApJ*, submitted (astro-ph/9907039).

DRELL, P. S., LOREDO, T. J. & WASSERMAN, I. 1999, *ApJ*, submitted (astro-ph/9905027).

RIESS, A. G., FILIPPENKO, A. V., LI, W. & SCHMIDT, B. P. 1999, *AJ*, submitted (astro-ph/9907038).

Jens Niemeyer, Jim Truran
University of Chicago
October, 1999

Acknowledgements

The editors wish to express their sincere appreciation to Carrie Clark, Stacy Cummings, and Mila Kuntu for their kind assistance with all aspects of the organization of the workshop. We thank Asantha Cooray for his help in the preparation of the manuscript for publication in Cambridge University Press. We thank the faculty and staff of the Department of Astronomy and Astrophysics and the Enrico Fermi Institute for their hospitality. One of us (JCN) wishes to acknowledge support of the Enrico Fermi Institute as a Fermi Fellow. We also acknowledge support by the ASCI Flash Center at the University of Chicago under DOE contract B341495.

Type Ia Supernovae and Their Cosmological Implications

By ALEXEI V. FILIPPENKO AND ADAM G. RIESS†

Department of Astronomy, University of California, Berkeley, CA 94720-3411 USA

We review the use of Type Ia supernovae (SNe Ia) for cosmological distance determinations. Low-redshift SNe Ia ($z \lesssim 0.1$) demonstrate that (a) the Hubble expansion is linear, (b) $H_0 = 65 \pm 2$ (statistical) km s^{-1} Mpc^{-1}, (c) the bulk motion of the Local Group is consistent with the COBE result, and (d) the properties of dust in other galaxies are similar to those of dust in the Milky Way. We find that the light curves of high-redshift ($z = 0.3$–1) SNe Ia are stretched in a manner consistent with the expansion of space; similarly, their spectra exhibit slower temporal evolution (by a factor of $1 + z$) than those of nearby SNe Ia. The luminosity distances of our first set of 16 high-redshift SNe Ia are, on average, 10–15% farther than expected in a low mass-density ($\Omega_M = 0.2$) universe without a cosmological constant. Preliminary analysis of our second set of 9 SNe Ia is consistent with this. Our work supports models with positive cosmological constant and a current acceleration of the expansion. We address many potential sources of systematic error; at present, none of them appears to reconcile the data with $\Omega_\Lambda = 0$ and $q_0 \geq 0$. The dynamical age of the Universe is estimated to be 14.2 ± 1.7 Gyr, consistent with the ages of globular star clusters.

1. Introduction

Supernovae (SNe) come in two main varieties (see Filippenko 1997b for a review). Those whose optical spectra exhibit hydrogen are classified as Type II, while hydrogen-deficient SNe are designated Type I. SNe I are further subdivided according to the appearance of the early-time spectrum: SNe Ia are characterized by strong absorption near 6150 Å (now attributed to Si II), SNe Ib lack this feature but instead show prominent He I lines, and SNe Ic have neither the Si II nor the He I lines. SNe Ia are believed to result from the thermonuclear disruption of carbon-oxygen white dwarfs, while SNe II come from core collapse in massive supergiant stars. The latter mechanism probably produces most SNe Ib/Ic as well, but the progenitor stars previously lost their outer layers of hydrogen or even helium.

It has long been recognized that SNe Ia may be very useful distance indicators for a number of reasons (Branch & Tammann 1992; Branch 1998, and references therein). (1) They are exceedingly luminous, with peak absolute blue magnitudes averaging -19.2 if the Hubble constant, H_0, is 65 km s^{-1} Mpc^{-1}. (2) "Normal" SNe Ia have small dispersion among their peak absolute magnitudes ($\sigma \lesssim 0.3$ mag). (3) Our understanding of the progenitors and explosion mechanism of SNe Ia is on a reasonably firm physical basis. (4) Little cosmic evolution is expected in the peak luminosities of SNe Ia, and it can be modeled. This makes SNe Ia superior to galaxies as distance indicators. (5) One can perform *local* tests of various possible complications and evolutionary effects by comparing nearby SNe Ia in different environments.

Research on SNe Ia in the 1990s has demonstrated their enormous potential as cosmological distance indicators. Although there are subtle effects that must indeed be taken into account, it appears that SNe Ia provide among the most accurate values of H_0, q_0 (the deceleration parameter), Ω_M (the matter density), and Ω_Λ (the cosmological constant, $\Lambda c^2 / 3 H_0^2$).

† On behalf of the High-z Supernova Search Team

There are now two major teams involved in the systematic investigation of high-redshift SNe Ia for cosmological purposes. The "Supernova Cosmology Project" (SCP) is led by Saul Perlmutter of the Lawrence Berkeley Laboratory, while the "High-Z Supernova Search Team" (HZT) is led by Brian Schmidt of the Mt. Stromlo and Siding Springs Observatories. One of us (A.V.F.) has worked with both teams, but his primary allegiance is now with the HZT. In this lecture we present results from the HZT.

2. Homogeneity and Heterogeneity

The traditional way in which SNe Ia have been used for cosmological distance determinations has been to assume that they are perfect "standard candles" and to compare their observed peak brightness with those of SNe Ia in galaxies whose distances have been independently determined (e.g., Cepheids). The rationale is that SNe Ia exhibit relatively little scatter in their peak blue luminosity ($\sigma_B \approx$ 0.4–0.5 mag; Branch & Miller 1993), and even less if "peculiar" or highly reddened objects are eliminated from consideration by using a color cut. Moreover, the optical spectra of SNe Ia are usually quite homogeneous, if care is taken to compare objects at similar times relative to maximum brightness (Riess et al. 1997, and references therein). Branch, Fisher, & Nugent (1993) estimate that over 80% of all SNe Ia discovered thus far are "normal."

From a Hubble diagram constructed with unreddened, moderately distant SNe Ia ($z \lesssim 0.1$) for which peculiar motions should be small and relative distances (as given by ratios of redshifts) are accurate, Vaughan et al. (1995) find that

$$< M_B(\text{max}) > = \ (-19.74 \pm 0.06) + 5\log(H_0/50) \text{ mag}. \tag{2.1}$$

In a series of papers, Sandage et al. (1996) and Saha et al. (1997) combine similar relations with *Hubble Space Telescope (HST)* Cepheid distances to the host galaxies of seven SNe Ia to derive $H_0 = 57 \pm 4$ km s^{-1} Mpc^{-1}.

Over the past decade it has become clear, however, that SNe Ia do *not* constitute a perfectly homogeneous subclass (e.g., Filippenko 1997a,b). In retrospect this should have been obvious: the Hubble diagram for SNe Ia exhibits scatter larger than the photometric errors, the dispersion actually *rises* when reddening corrections are applied (under the assumption that all SNe Ia have uniform, very blue intrinsic colors at maximum; van den Bergh & Pazder 1992; Sandage & Tammann 1993), and there are some significant outliers whose anomalous magnitudes cannot possibly be explained by extinction alone.

Spectroscopic and photometric peculiarities have been noted with increasing frequency in well-observed SNe Ia. A striking case is SN 1991T; its pre-maximum spectrum did not exhibit Si II or Ca II absorption lines, yet two months past maximum the spectrum was nearly indistinguishable from that of a classical SN Ia (Filippenko et al. 1992b; Phillips et al. 1992). The light curves of SN 1991T were slightly broader than the SN Ia template curves, and the object was probably somewhat more luminous than average at maximum. The reigning champion of well observed, peculiar SNe Ia is SN 1991bg (Filippenko et al. 1992a; Leibundgut et al. 1993; Turatto et al. 1996). At maximum brightness it was subluminous by 1.6 mag in V and 2.5 mag in B, its colors were intrinsically red, and its spectrum was peculiar (with a deep absorption trough due to Ti II). Moreover, the decline from maximum brightness was very steep, the I-band light curve did not exhibit a secondary maximum like normal SNe Ia, and the velocity of the ejecta was unusually low. The photometric heterogeneity among SNe Ia is well demonstrated by Suntzeff (1996) with five objects having excellent $BVRI$ light curves.

3. Cosmological Uses

3.1. *Luminosity Corrections and Nearby Supernovae*

Although SNe Ia can no longer be considered perfect "standard candles," they are still exceptionally useful for cosmological distance determinations. Excluding those of low luminosity (which are hard to find, especially at large distances), most SNe Ia are *nearly* standard (Branch, Fisher, & Nugent 1993). Also, after many tenuous suggestions (e.g., Pskovskii 1977, 1984; Branch 1981), convincing evidence has finally been found for a *correlation* between light-curve shape and luminosity. Phillips (1993) achieved this by quantifying the photometric differences among a set of nine well-observed SNe Ia using a parameter, $\Delta m_{15}(B)$, which measures the total drop (in B magnitudes) from maximum to $t = 15$ days after B maximum. In all cases the host galaxies of his SNe Ia have accurate relative distances from surface brightness fluctuations or from the Tully-Fisher relation. In B, the SNe Ia exhibit a total spread of ~ 2 mag in maximum luminosity, and the intrinsically bright SNe Ia clearly decline more slowly than dim ones. The range in absolute magnitude is smaller in V and I, making the correlation with $\Delta m_{15}(B)$ less steep than in B, but it is present nonetheless.

Using SNe Ia discovered during the Calán/Tololo survey ($z \lesssim 0.1$), Hamuy et al. (1995, 1996b) confirm and refine the Phillips (1993) correlation between $\Delta m_{15}(B)$ and $M_{max}(B, V)$: it is not as steep as had been claimed. Apparently the slope is steep only at low luminosities; thus, objects such as SN 1991bg skew the slope of the best-fitting single straight line. Hamuy et al. reduce the scatter in the Hubble diagram of normal, unreddened SNe Ia to only 0.17 mag in B and 0.14 mag in V; see also Tripp (1997).

In a similar effort, Riess, Press, & Kirshner (1995a) show that the luminosity of SNe Ia correlates with the detailed shape of the light curve, not just with its initial decline. They form a "training set" of light-curve shapes from 9 well-observed SNe Ia having known relative distances, including very peculiar objects (e.g., SN 1991bg). When the light curves of an independent sample of 13 SNe Ia (the Calán/Tololo survey) are analyzed with this set of basis vectors, the dispersion in the V-band Hubble diagram drops from 0.50 to 0.21 mag, and the Hubble constant rises from 53 ± 11 to 67 ± 7 km s^{-1} Mpc^{-1}, comparable to the conclusions of Hamuy et al. (1995, 1996b). About half of the rise in H_0 results from a change in the position of the "ridge line" defining the linear Hubble relation, and half is from a correction to the luminosity of some of the local calibrators which appear to be unusually luminous (e.g., SN 1972E).

By using light-curve shapes measured through several different filters, Riess, Press, & Kirshner (1996a) extend their analysis and objectively eliminate the effects of interstellar extinction: a SN Ia that has an unusually red $B - V$ color at maximum brightness is assumed to be *intrinsically* subluminous if its light curves rise and decline quickly, or of normal luminosity but significantly *reddened* if its light curves rise and decline slowly. With a set of 20 SNe Ia consisting of the Calán/Tololo sample and their own objects, Riess, Press, & Kirshner (1996a) show that the dispersion decreases from 0.52 mag to 0.12 mag after application of this "multi-color light curve shape" (MLCS) method. The results from a very recent, expanded set of nearly 50 SNe Ia indicate that the dispersion decreases from 0.44 mag to 0.15 mag (Riess et al. 1999b, in preparation). The resulting Hubble constant is 65 ± 2 (statistical) km s^{-1} Mpc^{-1}, with an additional systematic and zero-point uncertainty of ± 5 km s^{-1} Mpc^{-1}. Riess, Press, & Kirshner (1996a) also show that the Hubble flow is remarkably linear; indeed, SNe Ia now constitute the best evidence for linearity. Finally, they argue that the dust affecting SNe Ia is *not* of circumstellar origin, and show quantitatively that the extinction curve in external galaxies typically does not differ from that in the Milky Way (cf. Branch & Tammann 1992; but see Tripp 1998).

Riess, Press, & Kirshner (1995b) capitalize on another use of SNe Ia: determination of the Milky Way Galaxy's peculiar motion relative to the Hubble flow. They select galaxies whose distances were accurately determined from SNe Ia, and compare their observed recession velocities with those expected from the Hubble law alone. The speed and direction of the Galaxy's motion are consistent with what is found from COBE (Cosmic Background Explorer) studies of the microwave background, but not with the results of Lauer & Postman (1994).

The advantage of systematically correcting the luminosities of SNe Ia at high redshifts rather than trying to isolate "normal" ones seems clear in view of recent evidence that the luminosity of SNe Ia may be a function of stellar population. If the most luminous SNe Ia occur in young stellar populations (Hamuy et al. 1995, 1996a; Branch, Roman-ishin, & Baron 1996), then we might expect the mean peak luminosity of high-redshift SNe Ia to differ from that of a local sample. Alternatively, the use of Cepheids (Population I objects) to calibrate local SNe Ia can lead to a zero point that is too luminous. On the other hand, as long as the physics of SNe Ia is essentially the same in young stellar populations locally and at high redshift, we should be able to adopt the luminosity correction methods (photometric and spectroscopic) found from detailed studies of nearby samples of SNe Ia.

4. High-Redshift Supernovae

4.1. *The Search*

These same techniques can be applied to construct a Hubble diagram with high-redshift SNe, from which the value of q_0 can be determined. With enough objects spanning a range of redshifts, we can determine Ω_M and Ω_Λ independently (e.g., Goobar & Perlmutter 1995). Contours of peak apparent R-band magnitude for SNe Ia at two redshifts have different slopes in the Ω_M–Ω_Λ plane, and the regions of intersection provide the answers we seek.

Based on the pioneering work of Norgaard-Nielsen et al. (1989), whose goal was to find SNe in moderate-redshift clusters of galaxies, Perlmutter et al. (1997) and our HZT (Schmidt et al. 1998) devised a strategy that almost guarantees the discovery of many faint, distant SNe Ia on demand, during a predetermined set of nights. This "batch" approach to studying distant SNe allows follow-up spectroscopy and photometry to be *scheduled* in advance, resulting in a systematic study not possible with random discoveries. Most of the searched fields are equatorial, permitting follow-up from both hemispheres.

Our approach is simple in principle; see Schmidt et al. (1998) for details, and for a description of our first high-redshift SN Ia (SN 1995K). Pairs of first-epoch images are obtained with the CTIO or CFHT 4-m telescopes and wide-angle imaging cameras during the nights just after new moon, followed by second-epoch images 3–4 weeks later. (Pairs of images permit removal of cosmic rays, asteroids, and distant Kuiper-belt objects.) These are compared immediately using well-tested software, and new SN candidates are identified in the second-epoch images. Spectra are obtained as soon as possible after discovery to verify that the objects are SNe Ia and determine their redshifts. Each team has already found over 70 SNe in concentrated batches, as reported in numerous IAU Circulars (e.g., Perlmutter et al. 1995 — 11 SNe with $0.16 \lesssim z \lesssim 0.65$; Suntzeff et al. 1996 — 17 SNe with $0.09 \lesssim z \lesssim 0.84$).

Intensive photometry of the SNe Ia commences within a few days after procurement of the second-epoch images; it is continued throughout the ensuing and subsequent dark runs. In a few cases *HST* images are obtained. As expected, most of the discoveries are *on the rise or near maximum brightness*. When possible, the SNe are observed in filters which closely match the redshifted B and V bands; this way, the K-corrections become only a second-order effect (Kim, Goobar, & Perlmutter 1996). Custom-designed filters for redshifts centered on 0.35 and 0.45 are used by our HZT (Schmidt et al. 1998), when appropriate. We try to obtain excellent *multi-color* light curves, so that reddening and luminosity corrections can be applied (Riess, Press, & Kirshner 1996a; Hamuy et al. 1996a,b).

Although SNe in the magnitude range 22–22.5 can sometimes be spectroscopically confirmed with 4-m class telescopes, the signal-to-noise ratios are low, even after several hours of integration. Certainly Keck is required for the fainter objects (mag 22.5–24.5). With Keck, not only can we rapidly confirm a large number of candidate SNe, but we can search for peculiarities in the spectra that might indicate evolution of SNe Ia with redshift. Moreover, high-quality spectra allow us to measure the age of a supernova: we have developed a method for automatically comparing the spectrum of a SN Ia with a library of spectra corresponding to many different epochs in the development of SNe Ia (Riess et al. 1997). Our technique also has great practical utility at the telescope: we can determine the age of a SN "on the fly," within half an hour after obtaining its spectrum. This allows us to rapidly decide which SNe are best for subsequent photometric follow-up, and we immediately alert our collaborators on other telescopes.

4.2. *Results*

First, we note that the light curves of high-redshift SNe Ia are broader than those of nearby SNe Ia; the initial indications of Leibundgut et al. (1996) and Goldhaber et al. (1997) are amply confirmed with our larger samples. Quantitatively, the amount by which the light curves are "stretched" is consistent with a factor of $1 + z$, as expected if redshifts are produced by the expansion of space rather than by "tired light." We were also able to demonstrate this *spectroscopically* at the 2σ confidence level for a single object: the spectrum of SN 1996bj ($z = 0.57$) evolved more slowly than those of nearby SNe Ia, by a factor consistent with $1 + z$ (Riess et al. 1997). More recently, we have used observations of SN 1997ex ($z = 0.36$) at three epochs to conclusively verify the effects of time dilation: temporal changes in the spectra are slower than those of nearby SNe Ia by roughly the expected factor of 1.36 (Filippenko et al. 1999).

Following our Spring 1997 campaign, in which we found a SN with $z = 0.97$ (SN 1997ck), and for which we obtained *HST* follow-up of three SNe, we published our first substantial results concerning the density of the Universe (Garnavich et al. 1998a): $\Omega_M = 0.35 \pm 0.3$ under the *assumption* that $\Omega_{\text{total}} = 1$, or $\Omega_M = -0.1 \pm 0.5$ under the *assumption* that $\Omega_\Lambda = 0$. Our independent analysis of 10 SNe Ia using the "snapshot" distance method (with which conclusions are drawn from sparsely observed SNe Ia) gives quantitatively similar conclusions (Riess et al. 1998a).

Our next results, obtained from a total of 16 high-z SNe Ia, were announced at a conference in February 1998 (Filippenko & Riess 1998) and formally published in September 1998 (Riess et al. 1998b). The Hubble diagram (from a refined version of the MLCS method; Riess et al. 1998b) for the 10 best-observed high-z SNe Ia is given in Figure 1 (*Left*), while Figure 1 (*Right*) illustrates the derived confidence contours in the Ω_M–Ω_Λ plane. We confirm our previous suggestion that Ω_M is low. Even more exciting, however, is our conclusion that Ω_Λ is *nonzero* at the 3σ statistical confidence level. With the MLCS method applied to the full set of 16 SNe Ia, our formal results are $\Omega_M = 0.24 \pm 0.10$ if

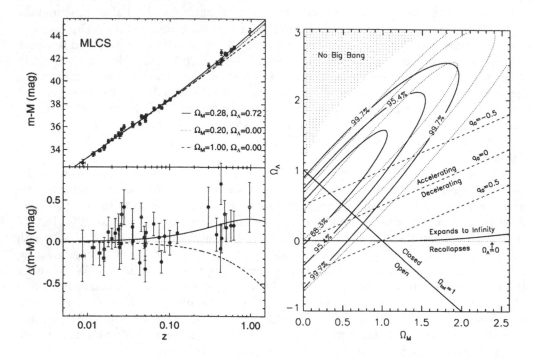

FIGURE 1. *Left:* The upper left panel shows the Hubble diagram for the low-redshift and high-redshift SN Ia samples with distances measured from the MLCS method; see Riess et al. (1998b). Overplotted are three world models: "low" and "high" Ω_M with $\Omega_\Lambda = 0$, and the best fit for a flat universe ($\Omega_M = 0.28$, $\Omega_\Lambda = 0.72$). The bottom left panel shows the difference between data and models from the $\Omega_M = 0.20$, $\Omega_\Lambda = 0$ prediction. Except for SN 1997ck (open symbol; $z = 0.97$), which lacks spectroscopic confirmation and was excluded from the fit, only the 9 best-observed high-redshift SNe Ia are shown. The average difference between the data and the $\Omega_M = 0.20$, $\Omega_\Lambda = 0$ prediction is 0.25 mag. *Right:* Joint confidence intervals for $(\Omega_M, \Omega_\Lambda)$ from SNe Ia (Riess et al. 1998b). The solid contours are results from the MLCS method applied to 10 well-observed SN Ia light curves, together with the snapshot method (Riess et al. 1998a) applied to 6 incomplete SN Ia light curves. The dotted contours are for the same objects excluding SN 1997ck ($z = 0.97$). Regions representing specific cosmological scenarios are illustrated.

$\Omega_{total} = 1$, or $\Omega_M = -0.35 \pm 0.18$ (unphysical) if $\Omega_\Lambda = 0$. If we demand that $\Omega_M = 0.2$, then the best value for Ω_Λ is 0.66 ± 0.21. These conclusions do not change significantly if only the 9 best-observed SNe Ia are used (Fig. 1; $\Omega_M = 0.28 \pm 0.10$ if $\Omega_{total} = 1$). The $\Delta m_{15}(B)$ method yields similar results; if anything, the case for a positive cosmological constant strengthens. (For brevity, in this paper we won't quote the $\Delta m_{15}(B)$ numbers; see Riess et al. 1998b for details.) From an essentially independent set of 42 high-z SNe Ia (only 2 objects in common), the SCP obtains almost identical results (Perlmutter et al. 1999). This suggests that neither team has made a large, simple blunder!

Very recently, we have calibrated an additional sample of 9 high-z SNe Ia, including several observed with *HST*. Preliminary analysis suggests that the new data are entirely consistent with the old results, thereby strengthening their statistical significance. Figure 2 (*Left*) shows the tentative Hubble diagram; full details will be published elsewhere.

Though not drawn in Figure 1 (*Right*), the expected confidence contours from measurements of the angular scale of the first Doppler peak of the cosmic microwave background radiation (CMBR) are nearly perpendicular to those provided by SNe Ia (e.g., Zaldar-

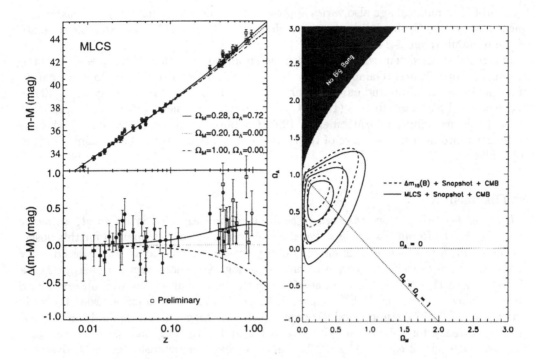

FIGURE 2. *Left:* As in Figure 1 (*Left*), the upper panel shows the Hubble diagram for the low-z and high-z SN Ia samples. Here, we include preliminary analysis of 9 additional SNe Ia (open squares). The bottom panel shows the difference between data and models from the $\Omega_M = 0.20$, $\Omega_\Lambda = 0$ prediction. *Right:* The HZT's combined constraints from SNe Ia (Fig. 1) and the position of the first Doppler peak of the CMB angular power spectrum; see Garnavich et al. (1998b). The contours mark the 68%, 95.4%, and 99.7% enclosed probability regions. Solid curves correspond to results from the MLCS method, while dotted ones are from the $\Delta m_{15}(B)$ method; all 16 SNe Ia in Riess et al. (1998b) were used.

riaga et al. 1997; Eisenstein, Hu, & Tegmark 1998); thus, the two techniques provide complementary information. The space-based CMBR experiments in the next decade (e.g., MAP, Planck) will give very narrow ellipses, but a stunning result is already provided by existing measurements (Hancock et al. 1998; Lineweaver & Barbosa 1998): our analysis of the data in Riess et al. (1998b) demonstrates that $\Omega_M + \Omega_\Lambda = 0.94 \pm 0.26$, when the SN and CMBR constraints are combined (Garnavich et al. 1998b; see also Lineweaver 1998, Efstathiou et al. 1999, and others). As shown in Figure 2 (*Right*), the confidence contours are nearly circular, instead of highly eccentric ellipses as in Figure 1 (*Right*). We eagerly look forward to future CMBR measurements of even greater precision.

The dynamical age of the Universe can be calculated from the cosmological parameters. In an empty Universe with no cosmological constant, the dynamical age is simply the "Hubble time" (i.e., the inverse of the Hubble constant); there is no deceleration. SNe Ia yield $H_0 = 65 \pm 2$ km s^{-1} Mpc^{-1} (statistical uncertainty only), and a Hubble time of 15.1 ± 0.5 Gyr. For a more complex cosmology, integrating the velocity of the expansion from the current epoch ($z = 0$) to the beginning ($z = \infty$) yields an expression for the dynamical age. As shown in detail by Riess et al. (1998b), we obtain a value of $14.2^{+1.0}_{-0.8}$ Gyr using the likely range for (Ω_M, Ω_Λ) that we measure. (The precision is so high because our experiment is sensitive to roughly the *difference* between Ω_M and

Ω_Λ, and the dynamical age also varies approximately this way.) Including the *systematic* uncertainty of the Cepheid distance scale, which may be up to 10%, a reasonable estimate of the dynamical age is 14.2 ± 1.7 Gyr.

This result is consistent with ages determined from various other techniques such as the cooling of white dwarfs (Galactic disk > 9.5 Gyr; Oswalt et al. 1996), radioactive dating of stars via the thorium and europium abundances (15.2 ± 3.7 Gyr; Cowan et al. 1997), and studies of globular clusters (10–15 Gyr, depending on whether *Hipparcos* parallaxes of Cepheids are adopted; Gratton et al. 1997; Chaboyer et al. 1998). Evidently, there is no longer a problem that the age of the oldest stars is greater than the dynamical age of the Universe.

5. Discussion

High-redshift SNe Ia are observed to be dimmer than expected in an empty Universe (i.e., $\Omega_M = 0$) with no cosmological constant. A cosmological explanation for this observation is that a positive vacuum energy density accelerates the expansion. Mass density in the Universe exacerbates this problem, requiring even more vacuum energy. For a Universe with $\Omega_M = 0.2$, the average MLCS distance moduli of the well-observed SNe are 0.25 mag larger (i.e., 12.5% greater distances) than the prediction from $\Omega_\Lambda = 0$. The average MLCS distance moduli are still 0.18 mag bigger than required for a 68.3% (1σ) consistency for a Universe with $\Omega_M = 0.2$ and without a cosmological constant. The derived value of q_0 is -0.75 ± 0.32, implying that the expansion of the Universe is accelerating. If Ω_Λ really is constant, then at least the region of the Universe we have observed ($z \lesssim 0.8$) will expand eternally. Under the simplifying assumption of global homogeneity and isotropy, the entire Universe will behave in this manner.

5.1. *Systematic Effects*

A very important point is that the dispersion in the peak luminosities of SNe Ia ($\sigma = 0.15$ mag) is low after application of the MLCS method of Riess et al. (1996a, 1998b). With 16 SNe Ia, our effective uncertainty is $0.15/4 \approx 0.04$ mag, less than the expected difference of 0.25 mag between universes with $\Omega_\Lambda = 0$ and 0.76 (and low Ω_M); see Figure 1 (*Left*). Systematic uncertainties of even 0.05 mag (e.g., in the extinction) are significant, and at 0.1 mag they dominate any decrease in statistical uncertainty gained with a larger sample of SNe Ia. Thus, our conclusions with only 16 SNe Ia are already limited by systematic uncertainties, *not* by statistical uncertainties — but of course the 9 new objects further strengthen our case.

Here we explore possible systematic effects that might invalidate our results. Of those that can be quantified at the present time, none appears to reconcile the data with $\Omega_\Lambda = 0$, though further work is necessary to verify this. Additional details can be found in Schmidt et al. (1998) and especially Riess et al. (1998b).

5.1.1. *Evolution*

The local sample of SNe Ia displays a weak correlation between light-curve shape (or luminosity) and host galaxy type, in the sense that the most luminous SNe Ia with the broadest light curves only occur in late-type galaxies. Both early-type and late-type galaxies provide hosts for dimmer SNe Ia with narrower light curves (Hamuy et al. 1996a). The mean luminosity difference for SNe Ia in late-type and early-type galaxies is ~ 0.3 mag. In addition, the SN Ia rate per unit luminosity is almost twice as high in late-type galaxies as in early-type galaxies at the present epoch (Cappellaro et al. 1997). These results may indicate an evolution of SNe Ia with progenitor age. Possibly relevant physical parameters are the mass, metallicity, and C/O ratio of the progenitor (Höflich, Wheeler, & Thielemann 1998).

We expect that the relation between light-curve shape and luminosity that applies to the range of stellar populations and progenitor ages encountered in the late-type and early-type hosts in our nearby sample should also be applicable to the range we encounter in our distant sample. In fact, the range of age for SN Ia progenitors in the nearby sample is likely to be *larger* than the change in mean progenitor age over the 4–6 Gyr lookback time to the high-z sample. Thus, to first order at least, our local sample should correct our distances for progenitor or age effects.

We can place empirical constraints on the effect that a change in the progenitor age would have on our SN Ia distances by comparing subsamples of low-redshift SNe Ia believed to arise from old and young progenitors. In the nearby sample, the mean difference between the distances for the early-type (8 SNe Ia) and late-type hosts (19 SNe Ia), at a given redshift, is 0.04 ± 0.07 mag from the MLCS method. This difference is consistent with zero. Even if the SN Ia progenitors evolved from one population at low redshift to the other at high redshift, we still would not explain the surplus in mean distance of 0.25 mag over the $\Omega_\Lambda = 0$ prediction.

Moreover, it is reassuring that initial comparisons of high-redshift SN Ia spectra appear remarkably similar to those observed at low redshift. For example, the spectral characteristics of SN 1998ai ($z = 0.49$) appear to be essentially indistinguishable from those of low-redshift SNe Ia; see Figure 3. In fact, the most obviously discrepant spectrum in this figure is the second one from the top, that of SN 1994B ($z = 0.09$); it is intentionally included as a "decoy" that illustrates the degree to which even the spectra of nearby SNe Ia can vary. Nevertheless, it is important to note that a dispersion in luminosity (perhaps 0.2 mag) exists even among the other, more normal SNe Ia shown in Figure 3; thus, our spectra of SN 1998ai and other high-redshift SNe Ia are not yet sufficiently good for independent, *precise* determinations of luminosity from spectral features (Nugent et al. 1995).

We expect that our local calibration will work well at eliminating any pernicious drift in the supernova distances between the local and distant samples. However, we need to be vigilant for changes in the properties of SNe Ia at significant lookback times. Our distance measurements could be especially sensitive to changes in the colors of SNe Ia for a given light-curve shape. Remember, our entire case for $\Omega_\Lambda > 0$ rests on a difference of only $\lesssim 0.25$ mag in apparent brightness from the $\Omega_\Lambda = 0$ ($\Omega_M = 0.20$) prediction!

5.1.2. *Extinction*

Our SN Ia distances have the important advantage of including corrections for interstellar extinction occurring in the host galaxy and the Milky Way. Extinction corrections based on the relation between SN Ia colors and luminosity improve distance precision for a sample of nearby SNe Ia that includes objects with substantial extinction (Riess, Press, & Kirshner 1996a); the scatter in the Hubble diagram is much reduced. Moreover, the consistency of the measured Hubble flow from SNe Ia with late-type and early-type hosts (see above) shows that the extinction corrections applied to dusty SNe Ia at low redshift do not alter the expansion rate from its value measured from SNe Ia in low dust environments.

In practice, our high-redshift SNe Ia appear to suffer negligible extinction; their $B - V$ colors at maximum brightness are normal, suggesting little color excess due to reddening. Riess, Press, & Kirshner (1996b) found indications that the Galactic ratios between selective absorption and color excess are similar for host galaxies in the nearby ($z \leq 0.1$) Hubble flow. Yet, what if these ratios changed with lookback time (e.g., Aguirre 1999a)?

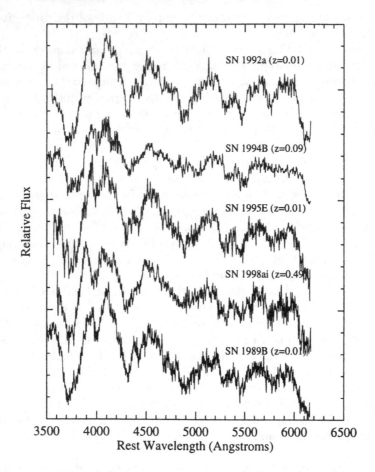

FIGURE 3. Spectral comparison (in f_λ) of SN 1998ai ($z = 0.49$; Keck spectrum) with low-redshift ($z < 0.1$) SNe Ia at a similar age ($\sim$ 5 days before maximum brightness), from Riess et al. (1998b) and Filippenko et al. (1999). The spectra of the low-redshift SNe Ia were resampled and convolved with Gaussian noise to match the quality of the spectrum of SN 1998ai. Overall, the agreement in the spectra is excellent, tentatively suggesting that distant SNe Ia are physically similar to nearby SNe Ia. SN 1994B ($z = 0.09$) differs the most from the others, and was included as a "decoy."

Could an evolution in dust grain size descending from ancestral interstellar "pebbles" at higher redshifts cause us to underestimate the extinction? Large dust grains would not imprint the reddening signature of typical interstellar extinction upon which our corrections rely.

However, viewing our SNe through such grey interstellar grains would also induce a *dispersion* in the derived distances. Using the results of Hatano, Branch, & Deaton (1998), Riess et al. (1998b) estimate that the expected dispersion would be 0.40 mag if the mean grey extinction were 0.25 mag (the value required to explain the measured MLCS distances without a cosmological constant). This is significantly larger than the 0.21 mag dispersion observed in the high-redshift MLCS distances. Furthermore, most of the observed scatter is already consistent with the estimated *statistical* errors, leaving little to be caused by grey extinction. Nevertheless, if we assumed that *all* of the observed scatter were due to grey extinction, the mean shift in the SN Ia distances would only be

0.05 mag. With the observations presented here, we cannot rule out this modest amount of grey interstellar extinction.

Grey *intergalactic* extinction could dim the SNe without either telltale reddening or dispersion, if all lines of sight to a given redshift had a similar column density of absorbing material. The component of the intergalactic medium with such uniform coverage corresponds to the gas clouds producing Lyman-α forest absorption at low redshifts. These clouds have individual H I column densities less than about $10^{15}\,\mathrm{cm}^{-2}$ (Bahcall et al. 1996). However, they display low metallicities, typically less than 10% of solar. Grey extinction would require larger dust grains which would need a larger mass in heavy elements than typical interstellar grain size distributions to achieve a given extinction. It is possible that large dust grains are blown out of galaxies by radiation pressure, and are therefore not associated with Lyman-α clouds (Aguirre 1999b). On the other hand, intergalactic grains reside in hard radiation environments that may be hostile to their survival. Finally, the existence of grey intergalactic extinction would only augment the already surprising excess of galaxies in high-redshift galaxy surveys (e.g., Huang et al. 1997).

We conclude that grey extinction seems like a somewhat implausible explanation for the observed faintness of high-redshift SNe Ia, but more work along the lines of Aguirre (1999b) is certainly warranted.

5.1.3. *Selection Bias*

Sample selection has the potential to distort the comparison of nearby and distant SNe. Most of our nearby ($z < 0.1$) sample of SNe Ia was gathered from the Calán/Tololo survey (Hamuy et al. 1993), which employed the blinking of photographic plates obtained at different epochs with Schmidt telescopes, and from less well-defined searches (Riess et al. 1999a). Our distant ($z \gtrsim 0.16$) sample was obtained by subtracting digital CCD images at different epochs with the same instrument setup.

Although selection effects could alter the ratio of intrinsically dim to bright SNe Ia in the nearby and distant samples, our use of the light-curve shape to determine the supernova's luminosity should correct most of this selection bias on our distance estimates. Nevertheless, the final dispersion is nonzero, and to investigate its consequences we used a Monte Carlo simulation; details are given by Riess et al. (1998b). The results are very encouraging, with recovered values of Ω_M or Ω_Λ exceeding the simulated values by only 0.02–0.03 for these two parameters considered separately. There are two reasons we find such a small selection bias in the recovered cosmological parameters. First, the small dispersion of our distance indicator ($\sigma \approx 0.15$ mag after light-curve shape correction) results in only a modest selection bias. Second, both nearby and distant samples include an excess of brighter than average SNe, so the *difference* in their individual selection biases remains small.

Additional work on quantifying the selection criteria of the nearby and distant samples is needed. Although the above simulation and others bode well for using SNe Ia to measure cosmological parameters, we must continue to be wary of subtle effects that might bias the comparison of SNe Ia near and far.

5.1.4. *Effect of a Local Void*

Zehavi et al. (1998) find that the SNe Ia out to 7000 km s^{-1} may (2–3σ confidence level) exhibit an expansion rate which is 6% greater than that measured for the more distant objects; see the low-redshift portion of Figure 1 (*Left*). The implication is that the volume out to this distance is underdense relative to the global mean density.

In principle, a local void would increase the expansion rate measured for our low-redshift sample relative to the true, global expansion rate. Mistaking this inflated rate for the global value would give the false impression of an increase in the low-redshift expansion rate relative to the high-redshift expansion rate. This outcome could be incorrectly attributed to the influence of a positive cosmological constant. In practice, only a small fraction of our nearby sample is within this local void, reducing its effect on the determination of the low-redshift expansion rate.

As a test of the effect of a local void on our constraints for the cosmological parameters, we reanalyzed the data discarding the seven SNe Ia within 7000 km s^{-1} ($d = 108$ Mpc for $H_0 = 65$ km s^{-1} Mpc^{-1}). The result was a reduction in the confidence that $\Omega_\Lambda > 0$ from 99.7% (3.0σ) to 98.3% (2.4σ) for the MLCS method.

5.1.5. *Weak Gravitational Lensing*

The magnification and demagnification of light by large-scale structure can alter the observed magnitudes of high-redshift SNe (Kantowski, Vaughan, & Branch 1995). The effect of weak gravitational lensing on our analysis has been quantified by Wambsganss et al. (1997) and summarized by Schmidt et al. (1998). SN Ia light will, on average, be demagnified by 0.5% at $z = 0.5$ and 1% at $z = 1$ in a Universe with a non-negligible cosmological constant. Although the sign of the effect is the same as the influence of a cosmological constant, the size of the effect is negligible.

Holz & Wald (1998) have calculated the weak lensing effects on supernova light from ordinary matter which is not smoothly distributed in galaxies but rather clumped into stars (i.e., dark matter contained in massive compact halo objects). With this scenario, microlensing becomes a more important effect, further decreasing the observed supernova luminosities at $z = 0.5$ by 0.02 mag for Ω_M=0.2 (Holz, private communication). Even if most ordinary matter were contained in compact objects, this effect would not be large enough to reconcile the SNe Ia distances with the influence of ordinary matter alone.

5.1.6. *Sample Contamination*

Riess et al. (1998b) consider in detail the possibility of sample contamination by SNe that are not SNe Ia. Of the 16 initial objects, 12 are clearly SNe Ia and 2 others are almost certainly SNe Ia (though the region near Si II λ6150 was poorly observed in the latter two). One object (SN 1997ck at $z = 0.97$) does not have a good spectrum, and another (SN 1996E) might be a SN Ic. A reanalysis with only the 14 most probable SNe Ia does not significantly alter our conclusions regarding a positive cosmological constant. However, without SN 1997ck we cannot obtain independent values for Ω_M and Ω_Λ.

Riess et al. (1998b) and Schmidt et al. (1998) discuss several other uncertainties (e.g., differences between fitting techniques; K-corrections) and suggest that they, too, are not serious in our study.

5.2. *Future Tests*

We are currently trying to increase the sample of high-z SNe Ia used for measurements of cosmological parameters, primarily to quantify systematic effects. Perhaps the most probable one is intrinsic evolution of SNe Ia with redshift: What if SNe Ia were *different* at high redshift than locally? One way to observationally address this is through careful, quantitative spectroscopy of the quality shown in Figure 3 for SN 1998ai. Nearby SNe Ia exhibit a range of spectral properties that correlate with light-curve shape, and we need to determine whether high-z SNe Ia show the same correlations. Comparisons of spectra obtained at very early times might be most illuminating: theoretical models predict that differences between SNe Ia should be most pronounced soon after the explosion, when the

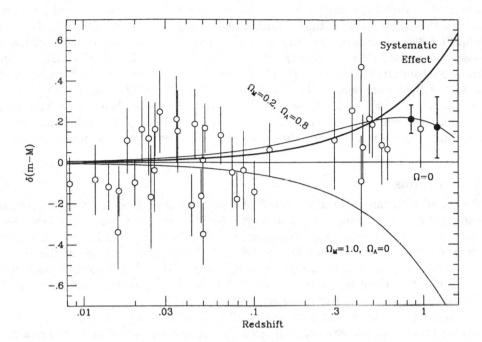

FIGURE 4. The HZT SN Ia data from Figure 1 (open circles) are plotted relative to an empty universe (horizontal line). The two faint curves are the best-fitting Λ model, and the $\Omega_M = 1$ ($\Omega_\Lambda = 0$) model. The darker curve shows a systematic bias that increases linearly with z and is consistent with our $z = 0.5$ data. The expected observational uncertainties of hypothetical SNe Ia at redshifts of 0.85 and 1.2 are shown (filled circles).

photosphere is in the outermost layers of the ejecta, which potentially undergo different amounts of nuclear burning.

Light curves can also be used to test for evolution of SNe Ia. For example, one can determine whether the rise time (from explosion to maximum brightness) is the same for high-z and low-z SNe Ia; a difference might indicate that the peak luminosities are also different (Höflich, Wheeler, & Thielemann 1998). Moreover, we should see whether high-z SNe Ia show a second maximum in the rest-frame I-band about 25 days after the first maximum, as do normal SNe Ia (e.g., Ford et al. 1993; Suntzeff 1996). Very subluminous SNe Ia do not show this second maximum, but rather follow a linear decline (Filippenko et al. 1992a).

One of the most decisive tests to distinguish between Λ and systematic effects utilizes SNe Ia at $z \gtrsim 0.85$. We observe that $z = 0.5$ SNe Ia are *fainter* than expected even in a non-decelerating universe, and interpret this as excessive distances (a consequence of positive Ω_Λ). If so, the deviation of apparent magnitude from the low-Ω_M, zero-Λ model should actually begin to *decrease* at $z \approx 0.85$; we are looking so far back in time that the Λ effect becomes small compared with Ω_M (see Fig. 4). If, on the other hand, a systematic effect such as non-standard extinction (Aguirre 1999a,b), evolution of the white dwarf progenitors (Höflich, Wheeler, & Thielemann 1998), or gravitational lensing (Wambsganss, Cen, & Ostriker 1998) is the culprit, we expect that the deviation of the apparent magnitude will continue growing, to first order (Fig. 4). Given the expected uncertainties, one or two objects clearly won't be sufficient, but a sample of 10–20 SNe Ia should give a good statistical result.

Note that a very large sample of SNe Ia in the range $z = 0.3$–1.5 could be used to see whether Λ varies with time ("quintessence" models — e.g., Caldwell, Dave, & Steinhardt 1998). This is a very ambitious plan, but it should be achievable within 5–10 years. Moreover, with a very large sample (200–1000) of high-redshift SNe Ia, and accurate photometry, it should be possible to conduct a "super-MACHO" experiment. That is, the light from some SNe Ia should be strongly amplified by the presence of intervening matter, while the majority will be deamplified (e.g., Holz 1998; Kantowski 1998). The distribution of amplification factors can be used to determine the type of dark matter most prevalent in the Universe (compact objects, or smoothly distributed).

6. Conclusions

When used with care, SNe Ia are excellent cosmological probes; the current precision of individual distance measurements is roughly 5–10%, but a number of subtleties must be taken into account to obtain reliable results. SNe Ia at $z \lesssim 0.1$ have been used to demonstrate that (a) the Hubble flow is definitely linear, (b) the value of the Hubble constant is 65 ± 2 (statistical) ± 5 (systematic) km s^{-1} Mpc^{-1}, (c) the bulk motion of the Local Group is consistent with that derived from COBE measurements, and (d) the dust in other galaxies is similar to Galactic dust.

More recently, we have used a sample of 16 high-redshift SNe Ia ($0.16 \leq z \leq 0.97$) to make a number of deductions, as follows. These are supported by our preliminary analysis of 9 additional high-z SNe Ia.

(1) The luminosity distances exceed the prediction of a low mass-density ($\Omega_M \approx 0.2$) Universe by about 0.25 mag. A cosmological explanation is provided by a positive cosmological constant at the 99.7% (3σ) confidence level, with the prior belief that $\Omega_M \geq 0$. We also find that the expansion of the Universe is currently accelerating ($q_0 \leq 0$, where $q_0 \equiv \Omega_M/2 - \Omega_\Lambda$).

(2) The dynamical age of the Universe is 14.2 ± 1.7 Gyr, including systematic uncertainties in the Cepheid distance scale used for the host galaxies of three nearby SNe Ia.

(3) These conclusions do not depend on inclusion of SN 1997ck ($z = 0.97$; uncertain classification and extinction), nor on which of two light-curve fitting methods is used to determine the SN Ia distances.

(4) The systematic uncertainties presented by grey extinction, sample selection bias, evolution, a local void, weak gravitational lensing, and sample contamination currently do not provide a convincing substitute for a positive cosmological constant. Further studies of these and other possible systematic uncertainties are needed to increase the confidence in our results.

We emphasize that the most recent results of the SCP (Perlmutter et al. 1999) are consistent with those of our HZT. This is reassuring — but other, completely independent methods are certainly needed to verify these conclusions. The upcoming space CMBR experiments hold the most promise in this regard. Although new questions are arising (e.g., What is the physical source of the cosmological constant, if nonzero? Are evolving cosmic scalar fields a better alternative?), we speculate that this may be the beginning of the "end game" in the quest for the values of Ω_M and Ω_Λ.

We thank all of our collaborators in the HZT for their contributions to this work. Our supernova research at U.C. Berkeley is supported by the Miller Institute for Basic Research in Science, by NSF grant AST-9417213, and by grant GO-7505 from the Space Telescope Science Institute, which is operated by the Association of Universities for Research in Astronomy, Inc., under NASA contract NAS 5-26555. A.V.F. and A.G.R.

are grateful for travel funds to this stimulating meeting in Chicago, from the Committee on Research (U.C. Berkeley) and the meeting organizers, respectively.

REFERENCES

AGUIRRE A. N. 1999a, ApJ, 512, L19

AGUIRRE A. N. 1999b, preprint (astro-ph/9904319).

BAHCALL J. N., et al. 1996, ApJ, 457, 19

BRANCH D. 1981, ApJ, 248, 1076

BRANCH D. 1998, ARAA 36 17.

BRANCH D., FISHER A., & NUGENT P. 1993, AJ, 106, 2383

BRANCH D., & MILLER D. L. 1993, ApJ, 405, L5

BRANCH D., ROMANISHIN W., & BARON E. 1996, ApJ, 465, 73; erratum 467, 473

BRANCH D., & TAMMANN G. A. 1992, ARAA, 30, 359

CALDWELL R., DAVE R., & STEINHARDT P. J. 1998, Phys. Rev. Lett., 80, 1582

CAPPELLARO E., et al. 1997, A&A, 322, 431

CHABOYER B., DEMARQUE P., KERNAN P. J., & KRAUSS L. M. 1998, ApJ, 494, 96

COWAN J. J., McWILLIAM A., SNEDEN C., & BURRIS D. L. 1997, ApJ, 480, 246

EFSTATHIOU G., et al. 1999, MNRAS, 303, L47

EISENSTEIN D. J., HU W., & TEGMARK M. 1998, ApJ, 504, L57

FILIPPENKO A. V. 1997a, in Thermonuclear Supernovae, Eds. P. Ruiz-Lapuente, et al., (Kluwer), 1

FILIPPENKO A. V. 1997b, ARAA, 35, 309

FILIPPENKO A. V., & RIESS A. G. 1998 Physics Reports, 307, 31

FILIPPENKO A. V., et al. 1992a, AJ, 104, 1543

FILIPPENKO A. V., et al. 1992b, ApJ, 384, L15

FORD C. H., et al. 1993, AJ, 106, 1101

GARNAVICH P., et al. 1998a, ApJ, 493, L53

GARNAVICH P., et al. 1998b, ApJ, 509, 74

GOLDHABER G., et al. 1997, in Thermonuclear Supernovae, Eds. P. Ruiz-Lapuente et al., (Kluwer), 777

GOOBAR A., & PERLMUTTER S. 1995, ApJ, 450, 14

GRATTON R. G., FUSI PECCI F., CARRETTA E., CLEMENTINI G., CORSI C. E., & LATTANZI M. 1997, ApJ, 491, 749

HAMUY M., et al. 1993, AJ, 106, 2392

HAMUY M., et al. 1995, AJ, 109, 1

HAMUY M., et al. 1996a, AJ, 112, 2391

HAMUY M., et al. 1996b, AJ, 112, 2398

HANCOCK S., ROCHA G., LAZENBY A. N., & GUTIÉRREZ C. M. 1998, MNRAS, 294, L1

HATANO K., BRANCH D., & DEATON J. 1998, ApJ, 502, 177

HÖFLICH P., WHEELER J. C., & THIELEMANN F. K. 1998, ApJ, 495, 617

HOLZ D. E. 1998, ApJ, 506, L1

HOLZ D. E., & WALD R. 1998, Phys. Rev. D, 58, 063501.

HUANG J. S., COWIE L. L., GARDNER J. P., HU E. M., SONGAILA A., & WAINSCOAT R. J. 1997, ApJ, 476, 12

KANTOWSKI R. 1998, ApJ, 507, 483

KANTOWSKI R., VAUGHAN T., & BRANCH D. 1995, ApJ, 447, 35

KIM A., GOOBAR A., & PERLMUTTER S. 1996, PASP, 108, 190

LAUER T., & POSTMAN M. 1994, ApJ, 425, 418

LEIBUNDGUT B., et al. 1993, AJ, 105, 301

LEIBUNDGUT B., et al. 1996, ApJ, 466, L21

LINEWEAVER C. H. 1998, ApJ, 505, L69

LINEWEAVER C. H., & BARBOSA D. 1998, ApJ, 496, 624

NORGAARD-NIELSEN H., et al. 1989, Nature, 339, 523

NUGENT P., PHILLIPS M., BARON E., BRANCH D., & HAUSCHILDT P. 1995, ApJ, 455, L147

OSWALT T. D., SMITH J. A., WOOD M. A., & HINTZEN P. 1996, Nature, 382, 692

PERLMUTTER S., et al. 1995, IAUC, 6270

PERLMUTTER S., et al. 1997, ApJ, 482, 565

PERLMUTTER S., et al. 1999, ApJ, in press

PHILLIPS M. M. 1993, ApJ, 413, L105

PHILLIPS M. M., et al. 1992, AJ, 103, 1632

PSKOVSKII YU. P. 1977, Sov. Astron., 21, 675

PSKOVSKII YU. P. 1984, Sov. Astron., 28, 658

RIESS A. G., NUGENT P. E., FILIPPENKO A. V., KIRSHNER R. P., & PERLMUTTER S. 1998a, ApJ, 504, 935

RIESS A. G., PRESS W. H., & KIRSHNER R. P. 1995a, ApJ, 438, L17

RIESS A. G., PRESS W. H., & KIRSHNER R. P. 1995b, ApJ, 445, L91

RIESS A. G., PRESS W. H., & KIRSHNER R. P. 1996a, ApJ, 473, 88

RIESS A. G., PRESS W. H., & KIRSHNER R. P. 1996b, ApJ, 473, 588

RIESS A. G., et al. 1997, AJ, 114, 722

RIESS A. G., et al. 1998b, AJ, 116, 1009

RIESS A. G., et al. 1999a, AJ, 117, 707

SAHA A., et al. 1997, ApJ, 486, 1

SANDAGE A., & TAMMANN G. A. 1993, ApJ, 415, 1

SANDAGE A., et al. 1996, ApJ, 460, L15

SCHMIDT B. P., et al. 1998, ApJ, 507, 46

SUNTZEFF N. 1996, in Supernovae and Supernova Remnants, Eds. R. McCray & Z. Wang, (Cambridge Univ. Press), 41

SUNTZEFF N., et al. 1996, IAUC, 6490

TRIPP R. 1997, A&A, 325, 871

TRIPP R. 1998, A&A, 331, 815

TURATTO M., et al. 1996, MNRAS, 283, 1

VAN DEN BERGH S., & PAZDER J. 1992, ApJ, 390, 34

VAUGHAN T. E., BRANCH D., MILLER D. L., & PERLMUTTER S. 1995, ApJ, 439, 558

WAMBSGANSS J., CEN R., & OSTRIKER J. P. 1998, ApJ, 494, 29

WAMBSGANSS J., CEN R. Y., XU G. H., & OSTRIKER J. P. 1997, ApJ, 475, L81

ZALDARRIAGA M., SPERGEL D. N., & SELJAK U. 1997, ApJ, 488, 1

ZEHAVI I., RIESS A. G., KIRSHNER R. P., & DEKEL A. 1998, ApJ, 503, 483

The Universe at $z \sim 1$: Implications for Type Ia Supernovae

By A. DRESSLER

Observatories of the Carnegie Institution, 813 Santa Barbara Street, Pasadena, CA 91101,
USA

Galaxy characteristics at $z \sim 1$, including morphology, stellar populations, and dust content, are reviewed. In comparison with the present properties of galaxies, possible implications for the nature of Type 1a supernovae at intermediate redshift are discussed.

1. Introduction

Type-Ia supernovae (SNe Ia) show great promise in the quest for the cosmological parameters of the universe. Perhaps they are the long sought "standard candle" (or, more appropriately, standard bomb) by which accurate distances, free from evolutionary effects and systematic errors, can be obtained. If so, we may be well on our way towards measuring the two remaining cosmological parameters Ω_o and Λ_o that characterize the Friedmann model.

Following in the tradition of Edwin Hubble, Allan Sandage made attempts in the 1960's and 70's to measure the deceleration of the universe, q_o, by observing the brightest galaxies in remote clusters with the Palomar Hale telescope. These efforts were frustrated by the realization that evolutionary effects of unknown sign and magnitude were likely to play as large a role as the geometry of the universe. The revival of this work in the last few years owes its great promise to the remarkably small dispersion found for SNe Ia, once a correction is made for the decline rate of a supernova's light curve. The small dispersion is crucial because the effect of different geometries is only a few tenths of a magnitude at $z \sim 0.5$, but an equally important feature of SNe Ia is the expectation that these explosive events are ruled by micro-physics (rather than the complicated evolution of a hundred-billion-solar-mass pile of gas, dust, and stars) so that their evolutionary changes might be negligible or, at the very least, predictable and correctable. As an added bonus, studies of supernovae of all types at cosmological distances will also be extremely valuable in understanding the history of star formation and heavy element enrichment over the age of the universe.

Type II supernovae arise from the core collapse of massive stars; this process is relatively well understood. As we heard at this meeting, a completely satisfactory model for type I supernovae has not been found, although it is likely that the basic elements – rapid nuclear burning of a carbon or oxygen white dwarf triggered by mass transfer from a companion star – are correct. The absence of a specific model will, however, limit our attempts to understand the systematics of SNe Ia, particularly the evolutionary corrections that might apply.

My purpose here is to describe some of the things we know about galaxies and stellar populations from $z \sim 1$ to the present, as a background for beginning to think about the possibility of evolutionary corrections that might apply to the brightness of SNe Ia. Certainly the great potential of this method, and the already provocative results from the two groups we will hear from, urge a careful scrutiny of these fantastic objects. I want to say right up front that I have no reason at this point to be suspicious of evolutionary corrections for these standard bombs; certainly one expects the physics of the explosive release of energy to be the same if the conditions are the same. The only doubt can come from concerns that there are properties of these objects that depend on the parts and how they combine, and how these might have changed subtlely over time as the typical stellar populations changed. As you will hear, tests can be made with present samples that test for effects due to age and metal abundance explore empirically, and so far there seems to be no cause for concern. And, eventually, we will have data that follow the predicted trajectory for a particular model in a redshift-magnitude diagram from z=0.5 to z=1.5 – this all by itself could provide powerful evidence that evolutionary effects are or are not important. But, for now, we are wise to adopt the motto "extraordinary claims require extraordinary proof" and look for potential troubles. For my own part, though, I would not be surprised, and certainly pleased, to learn someday that the effects that might have played no role at all!

2. Complications, complications

Our hope is that, since physics on a stellar scale is responsible for SNe Ia explosions, more global properties of the host galaxies, including stellar populations, environment, and cosmic evolution, will have minimal effect. Nevertheless, it's worth considering what are the kinds of things that are expected to vary with lookback time that could conceivably effect either the character of SNe Ia or our ability to observe them? Most of the candidates have to do with properties of the host galaxy, for example, heavy element abundances and star formation history. But, within these broad categories are specific things that should vary over time that might play a role. For example, heavy element abundance includes not just overall levels but such details as Fe/α-element abundances, and the formation history is not only a rate, but the initial mass function and the character of binarism, for example. Other details about star formation may effect the distribution of progenitor stars for SNe Ia, such as the degree of mass loss as a star ascends the giant branch, and the composition and mass function of sub-Chandrasekhar degenerate stars. Since we also expect gas and dust contents of galaxies to be, on average, systematically different in the past, the effect of dust extinction (both the amount and the reddening law associated with it) could influence our ability to find, observe, and interpret observations of distant supernovae.

We can be certain that some of these potential sources of variation do in fact change as we look back to $z \sim 1$. As I will now describe, there is evidence that star formation rates are higher at high redshift, and that starbursts are more common. Metal abundance does in fact build up with time, at different rates for different mass galaxies. The morphological types of host galaxies, which correlates with properties of the stellar population and star formation rate, is now known to evolve since $z \sim 1$, for galaxies in groups and clusters as well as those in the sparsely populated field. The most global characterizations of galaxies – luminosity and mass – appear to change slowly, in opposite directions, with the advance of cosmic time, and the global environment of galaxies includes both more grouping and clustering of galaxies and, simultaneously, the greater isolation of single galaxies. Such environmental differences no doubt result in changes in the merger and interaction rates for these systems.

While the preceding are things that certainly do change, the initial mass function, as a function of time, and the extinction due to dust – both the amount and the extinction law, are two properties that *might* have changed since $z \sim 1$, but for which we have no good evidence as yet.

3. Looking far back is getting to be routine

After decades of struggle to find typical galaxies beyond $z = 0.5$, which for many years represented the frontier of the Hubble diagram, observers now routinely survey the universe out to $z = 1$ and, with greater effort and stronger selection effects, even out to $z = 5$. The invention of the charge-coupled-device, and now the Hubble Space Telescope and new large telescopes like the Kecks, have made this possible. A stroll through the Hubble deep field will show that galaxies out to $z = 1$ are very recognizable: all the familiar Hubble types, from elliptical to irregular, are well represented half-a-Hubble-time ago. Our knowledge of galaxies $1 < z < 2.5$ is scant due to the redshift and the difficulty of working from the ground in the near-IR, and beyond $z = 2.5$ the appearance of galaxies is noticeably changed in both the size and form of galaxies.

However, while the universe at $z = 1$ may have familiar faces, there are some substantial changes. The large excess of faint blue galaxies over what is expected for non-evolving or passively evolving populations of today's galaxies (Glazebrook et al. 1994) has now been identified as due to a population of moderate luminosity (but probably low mass) dwarf galaxies, most of them with irregular morphology (Ellis 1997, and references therein). The strong decline of the global star formation rate (Lilly et al. 1996) since $z = 1$ is largely a reflection of the emergence and fading of these small but bright galaxies. Driver et al. (1995, 1996) have further shown that these galaxies have very irregular morphologies. It is probable, then, that any supernovae survey at $z = 1$ will contain a disproportionate number of young stars with relatively low metal abundances (Hammer et al. 1997) arising in these small systems seen in their prime, compared with a survey in the present epoch. Other, more detailed properties of the stellar populations, such as binarism, might also correlate.

For luminous galaxies like the Milky Way the situation is reassuring: evolution has been rather mild and is primarily limited to a higher star formation rate in spiral galaxies. The Canada-France Redshift Survey (Lilly et al. 1995) was an ambitious project to obtain photometry and redshifts for a complete sample of galaxies $17.5 < I_{ab} < 22.5$. Spectra of 591 of the 730 target galaxies show, by extrapolation, that the characteristic magnitude associated with a $z = 1$ galaxy is $23.5 < I_{ab} = 24.5$, corresponding to an R magnitude $R \sim 25$! (It is easy to understand, then, why no redshift survey has been done that typically samples $z = 1$ galaxies.) With this sample Lilly and collaborators are able to study the evolution of the luminosity function for galaxies in four redshift ranges up to $z = 1$. They find that red galaxies, associated with most of today's luminous galaxies, have undergone little luminosity or size evolution over this period in contrast with the rapidly evolving blue galaxies. Lilly et al (1998) and Brinchmann et al. (1998) have used HST observations to show that these redder galaxies are, indeed, associated with elliptical, S0, and spiral galaxies, with the strongly evolving blue population generally classified as irregular or peculiar. Lilly et al. have also shown that star formation rates at $z = 1$ are substantially elevated even in the luminous spiral population compared with the present epoch, with a corresponding rise in surface brightness and a blueing in $(U - V)_o$, but this is consistent with only a small change in the total luminosity since star formation rates are so low in most present epoch spirals. Lilly et al. find a best fitting model to the star formation history in $z \sim 1$ spirals as one with an e-folding time of 5 Gyr.

In summary, luminous galaxies were relatively similar to those we see today, albeit with elevated star formation rates, but that a much larger population of low-mass but luminous irregular galaxies was much more prominent. If the descendants of these latter systems are low surface brightness, dwarfish galaxies, it might be that a there are as many old stars in these systems today as in luminous galaxies. If so, one might expect to see many SNe Ia in relatively puny galaxies. I see little evidence for this in the pictures in Hamuy et al. (1997), but to my knowledge, the luminosity function of host galaxies for these objects has not been produced. If this is not seen, it suggests that the production of SNe Ia might well depend on heavy element abundance, which could be important for a systematic change in this population with increasing redshift.

4. Star formation histories from the fossil record

4.1. *Star formation rates back to z ∼ 1*

Even though we are learning much from observing star formation directly from lookback observations, most of what we know still comes from reading the fossil record of the stellar populations in our own and neighboring galaxies. In these cases we can resolve the stars down to the horizontal branch or down to the main sequence turnoff and to a remarkable extent dissect when generations of stars formed and their heavy element abundance. Much progress is now being made reading the history of star formation with these methods, especially in nearby dwarf galaxies.

A first comment is that the long controversy over the age spread of the globular clusters in the Milky Way halo seems to have been resolved in favor of the camp that believed that all the globulars are old, formed within one or two billion years of each other, some 12-15 billion years ago (Stetson et al. 1991). The implication that the Galactic halo is similarly old is consistent with the data, although there may be traces of younger populations strung through the halo, perhaps remnants of accretion events. The age distribution of stars in the Galactic bulge also seems to be generally old (Oertolani et al.) contrary to some earlier work that suggested late infall of gas and subsequent star formation.

Carrying this work to the spheroids of other galaxies, specifically elliptical galaxies, yields a similar conclusion, though from other methods relying on integrated stellar populations. Ellis et al. (1997) showed from a sample of elliptical galaxies in distant clusters that in this environment, at least, star formation in these systems must have occurred very early, $z > 3$ for the bulk of the stars. This added leverage to, and was consistent with, the results of Bower et al. (1992) for the Coma cluster. For elliptical galaxies in the field, the situation is not so clear; Gonzalez (1993) and Trager (1997) use high S/N measurements of line strength to argue that, while most of the stars in field ellipticals are old, the signs of recent star formation are found in a significant fraction of the sample.

Star formation in the disks of luminous galaxies, as was indicated by the CFRS study by Lilly et al., seem to have long histories stretching back far earlier than $z = 1$. The classic work by Twarog (1980) showing a declining star formation rate since 8 Gyr ago seems very consistent with this, as does subsequent work by Noh & Scalo (1989). The star formation picture in the Magellanic Clouds is noticeably different, with a more-or-less rising rate over the past 10^{10} years as judged from the cluster population (see, e.g., Alongi & Chiosi 1991), with an especially high star formation rate about 1 Gyr ago. Dwarf systems in the Local Group provide the most variety, with a whole range of star formation histories from old to young, with plenty of confined episodes of star formation along the way (see. e.g., Gallart et al. 1996 and Aparicio et al. 1997).

In summary, it seems to be not too much of a simplification to say that stars in luminous spheroids are generally old $z ∼ 2$, that the disks of luminous galaxies have a longer formation epoch, with high rates of star formation persisting to $z ∼ 1$ and declining rapidly thereafter, and that many small disk systems (including irregulars) actually formed at $z \sim 1$. The implication for SNe Ia is unclear, but assuming that they do arise from older stellar populations, it seems that the rate of SNe Ia per unit stellar mass could have been measurably different at $z \sim 1$ and that the progenitor population should have, in general, come from somewhat more massive stars stars than the do today. For most of these the heavy element abundance is likely to be only slightly lower, on average, than it is today. To the extent that stellar systems as young as a few Gyr with low heavy element abundance can develop SNe Ia, there could be a substantial increase in such systems.

4.2. *Heavy element evolution*

The production of the heavy elements is one of the fundamental processes in the universe. The overall level, isotopic abundances, and r-process compared to s-process elements show wide variation over the range of galactic environments, and through cosmic time.

The clearest record of the metal enrichment history of a luminous galaxy is derived for the Milky Way. Studies of the Galactic bulge (see, e.g. Rich 1990) show a wide range of 2 dex, ranging from $-1 < [Fe/H] < 1$ with a specific distribution that matches the "closed box model," that is, where all the gas is confined in the bulge and reused in succeeding generations of stars. The spheroids of other galaxies, i.e., ellipticals and large-buge spirals and S0's, are at least as tightly bound as our own, so it is reasonable to assume that they do have a wide range of abundances consistent with the closed box model, that is, with stars that are much more metal rich than the solar neighborhood but also a significant population of stars with $[Fe/H] \sim -1$. Since these populations are primarily old, this distribution is not expected to change much up to $z \sim 1$.

There has been much interesting work on the heavy element evolution in the Galactic disk in recent years. The surprising result is that, while the spread in abundance for stars forming today in the solar neighborhood is relatively small and near solar, the fossil record shows that at 5-10 Gyr ago the spread was much larger, ranging from one-fifth solar to solar (Edvardsson et al 1993). Apparently the disk was not well mixed at $z \sim 1$, perhaps indicative of the prolonged formation epoch discussed earlier.

It is well known that heavy element abundances in systems much smaller than the Milky Way are usually well below solar, only $[Fe/H] \sim -2$ for the smaller dwarf galaxies, much like the metal-poor population of the Galactic halo. Intermediate mass disk systems are probably well represented by the Magellanic clouds, which show a gradual enrichment from $[Fe/H] \sim -2$ to $[Fe/H] \sim -0.5$ over their history (Da Costa 1991).

Variation in chemical abundances occur not only in overall level but also in the relative abundance of the "α-elements," O, Mg, Ca, etc. As reviewed by McWilliam (1997), it is expected that α-elements will dominate over the first Gyr of a population because of their overproduction in Type II supernova in comparison to Type I, which "kick in" only after the lower mass range is reached to provide progenitors. Indeed, the Galactic bulge population shows a clear trend that as the metal abundance rose from $[Fe/H] \sim -1$ to solar, that is, as the population was enriched by SNe Ia events, the enhancement of α-elements, approximately a factor of two over the solar level, was gradually diluted until it reached the solar value. For the Galactic disk the situation is not as clear, with Si and Ca following the expected dilution, but Mg, and Ti abundances remaining constant over time. These observations show how complex the chemical enrichment history of the disk must have been, and how much is left to be learned from a careful study of individual elements in a wide range of stellar types and ages.

In summary, the range of chemical abundance in halos and spheroids of luminous galaxies has probably not changed much since $z \sim 1$. Abundances of disks should vary considerably, with lower overall levels and variations in specific abundances compared to the present at the earlier epoch. Dwarf galaxies probably represent an intermediate case, with a similar upper bound of $-1 < [Fe/H] < -2$ as it is today, but probably a greater number of systems with much lower metal abundance, experiencing their first significant epoch of star formation.

4.3. *The character of star formation*

The rate of star formation and the heavy element abundance are two characteristics of a stellar population, but they are not the only ones that may be important in the present context. Chief among those is the possible variation and/or evolution of the initial mass function (IMF). Years of work has consistently validated the ubiquity of the Salpeter power-law mass function, with a slope of -2.3. Surprisingly little variation is found over a wide range of environments, although a difficulty is that only relatively a small range in the mass function is explored in each study. For example, a study of Galactic open clusters by Elizabeth Lada (1998, private communication) reports no measurable variation on the the standard IMF over a range of several orders of magnitude in clusters richness (total mass). On the other hand, Phil Massey's study of the upper end of the IMF in the Large Magellanic Cloud shows a steepening (more intermediate-mass stars per high-mass star) for stars that are found in very poor clusters, especially for a population of apparently isolated O and B stars (Massey et al. 1995, Massey 1998) Interestingly, this might correspond to suggestive evidence that the IMF is flatter in starburst systems (e.g., Arp 220, M82, see Heckman 1998). Even in high redshift clusters, starburst galaxies seem to be using up an implausible amount of gas in star formation if the IMF is follows a Salpeter function down to M stars. Perhaps the IMF does change in the sense that relatively more massive stars are made when a great deal of raw material is around.

There is no evidence that the IMF depends on heavy element abundance, but, in truth there is little evidence one way or the other. Learning about this and other sensitivities of the IMF are difficult programs, but without this knowledge it will remain impossible to predict systematic changes in the IMF with cosmic time, leaving us with only empirical tests of SNe Ia samples to rule out this potential problem.

Finally, one wonders if the incidence of binarism depends on the IMF or heavy element abundance. A study by Carney & Latham (1987) of halo binaries suggest that heavy element abundance is not an important factor, but again the data are sparse and there is virtually no data on binarism as a function of the IMF or even simply of mass. These are two more factors that may be different at $z \sim 1$ that may systematically change the progenitor populations for SNe Ia if, as we suspect, binaries including evolved stars are the key components.

5. Dusts and Starbursts, and Dusty Starbursts

Finally, it is important to take note of two factors that could be undergoing substantial evolution of the redshift range $0 < z < 1$. One is the presence of dust. We know that intergalactic dust, if it exists, does not play any role in extinguishing the light from distant quasars up to redshift $z = 3$. However, the prevalence of dust in galaxies may have changed substantially even since $z \sim 1$. The strong evolution of ultraluminous infrared galaxies is one possible clue, but these are far from ordinary systems. More applicable is the first detections by SCUBA on UKIRT of sub-millimeter radiation from distant galaxies, light that arises as a direct result of dust re-radiation in rapidly star-forming regions (see, e.g., Blaine et al 1998). The possibility of greater dust absorption also is indicated by a comparison of the NICMOS images of the Hubble Deep Field (Thompson et al. 1999): some galaxies "fill in" in the infrared picture (actually rest frame R and I for $z \sim 1$) compared to the original visible-light HDF (see also results from ISO, Rowan-Robinson et al. 1997) This is good evidence for dust in many cases, although it is also clear that the majority of galaxies do not show any substantial effect. Increasingly more sensitive sub-mm observations, and the coming Space Infrared Telescope Facility, will shed a lot of light on this subject, so to speak. The implications for SNe Ia at high redshift are obvious, not only in the amount of extinction but also possible variations in reddening law, which is at this point hardly more than a subject for speculation.

My own work with the "MORPHS" collaboration, a study of intermediate-redshift ($z \sim 0.5$) clusters with HST and ground-based photometry and spectroscopy, has added an interesting datum. For many years it has been known that starbursts are far more numerous in these dense environments than they are today (Dressler & Gunn 1983, Dressler et al. 1992). This is derived from the large number of galaxies whose spectra match a post-starburst phase. The cause is still unknown, but it is likely that some combination of tidal and ram-pressure from a hot intergalactic medium is responsible. However, what has also been puzzling, for just as long, is the lack of progenitors for these systems: far too few luminous starbursts are caught "in the act." The MORPHS advance the explanation that another well populated class, also with post-starburst features but with a healthy rate of continuing star formation, are in fact dusty starbursts in progress with most of their young stars hidden from view by while the older A stars have diffused out of the strongly obscured star forming regions (Poggianti et al. 1999). (Galaxies of this type are found, though rare, at the present epoch and are confirmed to be dust laden by their copious far-IR emission.) Since MORPHS study finds an comparably large fraction of this latter type in the intermediate-redshift field, the implication is that starbursts, particularly dusty ones, are far more common even at intermediate redshifts. Confirmation will require detection of their far-IR radiation, but, if true, could signal a significant increase in uncommon but not at all rare dusty starbursting galaxies at $z \leq 0.5$.

6. Summary: What to expect for $z \sim 1$ galaxies

The following are what I consider to be likely characteristics of the galaxy population at $z \sim 1$ that may be relevant for studies of distant supernovae:

• Many more low mass, but high luminosity galaxies, presumably metal poor.

• A population of massive galaxies similar to today, with little change in their spheroids and several times greater star formation rates in their disks. Possibly a wider spread in metal abundances than is seen in today's disks.

• A systematic decline in heavy element abundance with increasing redshift. Many more stars with substantially less than solar abundance

• Probably a higher fraction of stars made in starbursts, with the possibility of a flatter IMF for these.

• More dust per galaxy, possibly a lot more in many galaxies, perhaps with extinction laws different from our own Galaxy.

• Fewer rich clusters of galaxies (high density environments) but more numerous low-velocity encounter and mergers in galaxy groups (not discussed here).

• Generally, structure in $z \sim 1$ galaxies will be less mature, more irregular, and more subject to secular change and sudden disturbance.

• Possible change in frequency or characteristics of binary star systems

REFERENCES

ALONGI M., & CHIOSI C. 1991, in The Magellanic Clouds, Eds. R. Haynes & D. Milne, (Kluwer) 193

APARICIO A., GALLART C., & BERTELLI G. 1997, AJ, 114, 680

BOWER R.G., LUCEY J.R., & ELLIS R.S. 1992, MNRAS, 254, 601

BLAINE A.W., SMAIL I., IVISON R.J.. & KNEIB J.-P. 1999, MNRAS, 302, 632

BRINCHMANN J., SCHADE D., TRESSE L., ELLIS R.S., LILLY S., LE FEBRE O., GLAZEBROOK K, HAMMER F., COLLESS M., CRAMPTON D., & BROADHURST T. 1998, ApJ, 449, 112

CARNEY B.S., & LATHAM D..W. 1987, AJ, 93, 116

DRIVER S.P., WINDHORST R.A., OSTRANDER E.J., KEEL W.C., GRIFFITHS R.E., & RATNATUNGA K.U. 1995, ApJ, 449, L23

DRIVER S.P., WINDHORST R.A., & GRIFFITHS R.E. 1996, in New Light on Galaxy Evolution, Eds. R. Bender & R.L. Davies, (Kluwer) 221

DA COSTA G.S. 1991, in The Magellanic Clouds, Eds. R. Haynes & D. Milne, (Kluwer) 183

DRESSLER A., & GUNN J.E. 1983, ApJ, 270, 7

DRESSLER A., SMAIL I., POGGIANTI B.M., BUTCHER H., COUCH W.C., ELLIS R.S., & OEMLER A.JR. 1999, ApJ, in press

EDVARDSSON B. GUSTAFSSON B., LAMBERT D.L., NISSEN P.E., & TOMKIN J. 1993, A&A, 275, 101

ELLIS R.S. 1997, ARAA, 35, 389

ELLIS R.S., SMAIL I., DRESSLER A., COUCH W.J., OEMLER A., BUTCHER H., & SHARPLES R.M. 1997, ApJ, 483, 582

GALLART C., APARICIO A., BERTELLIS G., & CHIOSI C. 1996, AJ, 112, 1950

GLAZEBROOK K., PEACOCK J.A., COLLINS C.A., & MILLIER L. 1994, MNRAS, 266, 65

GONZALEZ J.J. 1993, Ph.D. Thesis, University of California, Santa Cruz

HAMMER F., FLORES H., LILLY S.J., CRAMPTON D., LE FEVRE O., ROLA C., MALLEN-ORNELAS G., SCHADE D., & TRESSE L. 1997, ApJ, 481, 49

HAMUY M. ET AL. 1997, AJ, 112, 2408

HECKMAN M.R. 1998, in Origins, Eds. C.E.Woodwar, J.M. Shull & H.A. Thronson, Jr. (ASP) 27

LILLY S.J., TRESSE L., HAMMER F., CRAMPTON D., & LE FEVRE O. 1995, ApJ, 455, 108

LILLY S., LE FEVRE O., HAMMER F., & CRAMPTON D. 1996, ApJ, 460, L1

LILLY S., & ET AL. 1998, ApJ, 500, 75

MASSEY P., LANG C.C., DeGIOIA-EASTWOOD K., & GARMANY C.D. 1995, ApJ, 438, 188

Massey P. 1998, in The Stellar initial Mass Function Eds. G. Gilmore & D. Howell, (ASP) 17

McWILLIAM A. 1997, ARAA, 35, 503

NOH H-R. AND SCALO J. 1989, ApJ, 352, 605

RICH M.R. 1990, ApJ, 362, 604

ROWAN-ROBINSON M., ET AL. 1997, MNRAS, 289, 490

STETSON P.B., VANDENBERG D.A., & BOLTE M., 1991, PASP, 108, 560

THOMPSON R.I., STORRIE-LOMBARDI L.J., WEYMANN R.J., RIEKE M.J SCHNEIDER G., STOBIE E., & LYTLE D. 1998, AJ, 117, 17

TRAGER S.C. 1997, Ph.D. Thesis, University of California, Santa Cruz

TWAROG B.A. 1980, ApJ, 242, 242

Probable Effects of Interstellar Extinction on Searches for SNe at $z \sim 1$

By DONALD G. YORK

Departments of Astronomy & Astrophysics, Enrico Fermi Institute
The University of Chicago, Chicago, IL 60637-1433 USA

The source of interstellar extinction, the so called reddening of starlight by interstellar consitutents, is not known. But recent empirical results suggest that the extinction per hydrogen atom decreases as redshift increases. Studies of supernovas at higher and higher redshift should thus not need large and uncertain corrections for global extinction, though a way to independently prove this for each sight-line should be found. A few suggestions are made, related to element abundances and to diffuse interstellar bands.

1. Introduction

Studies of the intrinsic luminosities of any objects at high redshift are subject to the differential effects of extinction, either from our Galaxy, from an intervening galaxy or from the distant region in which the objects occur. (I ignore here the effects of extinction within our Galaxy, as the effect is well known and there are various ways to make corrections for it). Unfortunately, while we have some empirical knowledge of extinction and how it changes from place to place, we have little understanding of the source of the extinction or why it displays different properties in different places. Thus, a priori correction of luminosities of high redshift supernovas, in particular, cannot be made. However, there are empirical reasons to suspect that the effects will not be large, for either searches or analysis. We review those reasons below herein.

2. Background

2.1. *History of the Postulate of Extinction*

Chesaux (1744) may have been the first with cause to wonder about general extinction (space filling and uniform in all directions). That was his own explanation of the dilemma of the darkness of the night sky, later known as Olbers' paradox (Harrison 1987). Olbers argued for a uniform extinction as well (Olbers 1823), and used Zodiacal light and comet tails as proof that particles existed in space adequate to justify the hypothesis. Heber Curtis seems to have been one of the first to place William Herschel's "holes in the sky" (dark nebulae) (Crowe 1994) into context and express the view that dust was patchily distributed throughout space, allowing him to explain the apparent correlation between galaxy counts and angle with respect to the plane of the Milky Way (Curtis 1921), which correlation greatly confused the then current debate on the nature of the nebulae.

2.2. *Nature of the Extinguishing Material*

Interstellar extinction is thought to be a combination of scattering, interference and absorption effects by small solid particles and/or large molecules. A systematic relationship between the diameter of Galactic open clusters and the apparent magnitude of the stars in each cluster gave the first definite clue as to the importance of some interstellar constituent (Trumpler 1930). The reddening effect noted by Trumpler was quantified at the end of the 1930's (Stebbins et al. 1940). Spitzer (1948) showed that the extinction was associated with star formation. Van de Hulst (1957) and Greenberg (1968) studied the effects of small particles thru Mie scattering, and such an explanation is the most common now discussed. While studies of interstellar polarization, in particular, the wavelength dependence and a sign inversion near 6000 Å, show that there certainly are particles in space (Aannestad & Purcell 1973), there is no sure argument that there are enough of them to cause all of the observed extinction. An argument by Field (1974) suggesting the red giants could produce solid grains in adequate number was later withdrawn (Field 1975) in favor of star forming nebulae.

Moreover, in 1956, Platt (1956) suggested that large molecules might have some influence on the extinction and might exist in interstellar space. At least one absorption effect empirically associated with the extinction, the diffuse interstellar bands (Herbig 1993, 1995), seems now most likely to be a molecular phenomenon. Internal conversion in large molecules was suggested by Douglas (Douglas 1977) and Smith et al. (1977) to be the most likely mechanism for forming the bands. (There are over 100 diffuse , i.e., non-atomic, bands from 4000 Å to 10,000 Å). Long chain carbon molecules in space are known (Bell et al. 1997; Broten et al. 1978) (C_3, HC_nN, where $n = 3$-11). According to McCarthy et al. (1997), it is likely that larger molecules of the same type exist, as there is not a dramatic dropoff in column density per H atom as the subscript, n, increases. In this regard, we note that that the newly detected seed molecule H_3^+ has a column density that is not directly related to number of H atoms either (McCall et al. 1998), being of substantially the same column density toward the star VI Cygni No. 12, [E(B-V)~3.3] as toward the young stellar ojbect GL2136 [E(B-V)> 10]. Recent suggestions that long negative carbon chains may be responsible for the diffuse interstellar bands have been made (Tulej et al. 1998ab). Aromatic polycyclic hydrocarbons are often said to be responsible for broad features generally related to the extinction seen between 3 and 11 microns, though the specificity of the features is much lower than for the optical bands (Donn 1968; Salama et al. 1996).

2.3. *Arguments Against Significant Extinction Effects*

Given all of the possibilities, it is necessary to appeal to general arguments to address the topic of this paper. In what follows, we use a general argument attributable to Purcell (1969), a set of now very specific observations of missing interstellar atoms (inferred to be locked in "grains") and the general decrease in abundances with redshift in QSO absorption lines to suggest that corrections for extinction of SNe at redshifts above 0.5 will be smaller and smaller as redshift increases. Any extinction that does matter will be from material in the vicinity of the SNe itself, not from the accumulation of material in intergalactic space. Features in the interstellar extinction curve would need to be observed to prove substantial extinction in the galaxy local to the SNe. Depletion patterns in intervening interstellar material, along the lines of studies noted below, could also be used, but these require high resolution spectra which will be hard to obtain for the high z supernovas or the galaxies in which the SNe arise.

3. Determining Extinction

3.1. *Fundamental limits*

Purcell showed that if the amount of extinction could be specified, then the mass of the material involved, per H atom, could be derived. One can worry that the way of specifying the extinction is by the differential color effects. "Snowballs", larger than the size of grains that produce differential extinction (effectively, larger than 2000 Å in diameter), would produce gray extinction that could not be recognized except by its effect on the luminosity of objects at known distances. For the use of any objects thought to be standard candles, a circular argument would be involved, since the distance is not known a priori. Purcell dealt with the conventional interpretation of the extinction curve as being fully reflected by differential terms, with the optical absorption, $A_v \sim 3 \times E(B\text{-}V)$, where E(B-V) is the excess of extinction in the B band (4400 Å) over that in the V band (5500 Å).

The extinction requires a mass in baryons of 2×10^{-3} times the mass of protons on the line of sight (Spitzer 1978).

3.2. *Carbon only?*

Reference to well determined solar or cosmic abundances (Grevesse & Anders 1998) reveals that the only elements capable of providing such a large mass are carbon and oxygen. The well-known high depletions of other elements: Fe, Si, Mg, Al, Ti, Ca etc., cannot account for this much extinction. Some carbon must be invoked. Oxygen is ruled out because it is empirically known to be undepleted (Meyer et al. 98). Carbon, on the other hand, may be depleted by a factor of two in diffuse interstellar clouds. Hobbs et al. (1982) showed this result toward the star δ Sco. It has since been verified by using the same forbidden transition of C II (λ 2326 Å) toward other stars (Sofia et al. 1997), as well as by the behavior of C I (Jenkins & Shaya 1979). Note that in the case of diffuse clouds, the total amount of C tied up in molecular form in known molecules is less than 1% of the total. Either the grains or large molecules (not yet proven to exist in diffuse clouds) must contain the missing carbon. It is now known that carbon is uniformly depleted in all types of interstellar gas, as if the resulting extinction is uniform per H atom and does not change in character between translucent clouds and more diffuse clouds (Sofia et al. 1997).

3.3. *Other elements*

Welty et al. (1997) have summarized the latest information on depletion of elements other than C, O and N, based on the studies done with the *Copernicus* satellite as well as the numerous studies using the *Hubble Space Telescope* (Meyer et al. 1998; Sofia et al. 1994; Cardelli 1994; Hobbs et al. 1993). Large amounts of specific species (Fe, Si, Al, Ca, Ti and others) are highly depleted in neutral clouds near the disk of the Milky Way. These results are consistent with the statements in the above section.

4. Comparison of Extinction and Element Depletions in the Local Group and in QSO Absorbers

4.1. *Extinction*

Several authors (Ostriker & Heisler 1984; Pei et al. 1991) have suggested that the decreasing number of QSOs per unit volume above $z = 2$ is due to dust in intervening galaxies. Ostriker et al. (1990) showed that if the extinction curve in the SMC is taken as a model, the effects of extinction would be small for redshifts $z < 3$, and that the abundance of all the elements (depleted into interstellar extinguishing particles or not) is so low by that redshift that extinction is not likely to be important. The extinction curve in the SMC is steeper than that of our Galaxy but has a reference visual extinction per H atom only 1/10 that seen in the Milky Way. The 2175 Å excess in Galactic extinction ("the 2200 Å bump") is missing in most cases. The LMC has an extinction effect intermediate between that of the Galaxy and the SMC (Mathis 1990). (It should be noted here that extinction curves measure extinction in reddened stars with reference to unreddened stars of the same spectral type. There are possible errors in this procedure:

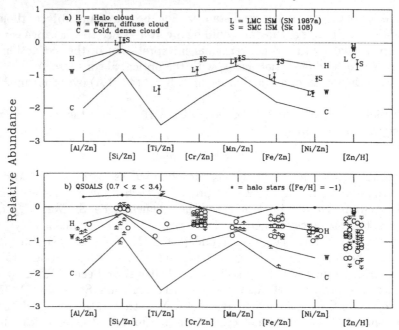

FIGURE 1. Average abundances (solid or open circles) for systems of interstellar lines in the LMC/SMC (top) or QSO absorption lines (bottom). Solid lines connect mean abundances for three distinct types of interstellar clouds. Abundances are referenced to zinc (Zn II) to minimize ionization corrections for the first ions observed for the other elements shown. At the right is an indicator of total metallicity $[Zn/H]$ based on $[ZnII/HI] - [Zn/H]_\odot$.

the latter stars are invariably much brighter, there could be effects due to scattered light from nebulosity near either type of star, and the spectral matching is never perfect. However, none of these effects seem to dominate the extinction itself.)

4.2. *Depletions*

The use of the *International Ultraviolet Explorer* (IUE) and the *Hubble Space Telescope* (HST) has allowed the extension of Galactic elemental gas depletion studies to the SMC and LMC. Using new data from HST for the star Sk 108 and ground based, very high resolution data on the species Ca^+, Welty et al. (1997) concluded that the species depleted in the interstellar regions of our Galaxy do not appear to be depleted by much at all in the SMC. Using IUE data on SN 1987A and very high resolution data on Na^o, K^o and Ca^+, toward the same object Welty et al. (1999) showed that while not depleted as much as gas in our Galaxy, some regions of the LMC are more depleted than most of the SMC gas. Combining these results with all published data on abundances in QSO absorption line systems for which H I is well determined (damped Lyman alpha systems), Welty et al. (1997) showed that the latter resemble the SMC, to the exclusion of LMC or Galactic-like depletion patterns. The general result is shown in Figure 1. For our purposes, the systematic errors are lower if one uses Zn^+ as the reference element (thus introducing only minimal ionization corrections, as the other elements are similarly observed as first ions). The absolute abundances ([Zn/H]) are shown at the far right. While the data do not reflect the pattern of element abundances from nucleosynthesis by a pure population of SNe II as might be expected at such high redshifts (Timmes et al. 1995), it is clear that depletions as extreme as seen in our Galaxy have not yet been encountered.

The intervening galaxies that could affect a statistical sample of SNe at high redshift are taken here to be the same types of objects that lead to QSO absoption systems. The observational definition in this case is that the intervening galaxy is not seen directly on top of the QSO, either because the QSO is far from the main galaxy or because the main galaxy has low surface brightness or is intrinsically faint. A fully formed L_* spiral lying near to or overlaying a distant SNe should immediately lead to rejection of the SNe from the sample. The existence of such a situation probably excludes the detection of that distant SNe anyway.

If intervening galaxies at high z are anything like galaxies at low z, especially with regard to the ISM, one can argue that dust clouds that cause sensible extinction are physically small. The chance of a coincidence between a small cloud and a QSO is small. The QSO test can be thought of as probing the intercloud medium.

In our Galaxy, with known highly depleted clouds, the intercloud medium contains depleted gas as well (York & Kinahan 1979) (the "warm" gas of Figure 1). For low velocity clouds, all sight lines near the Sun show this effect, so any QSO seen by a distant observer through our Galaxy would intersect the intercloud medium and show depletion, even though a translucent cloud with high depletion were not intersected. We can conclude that observations of QSO absorption lines would have revealed the existence of intercloud media with Milky Way type depletion if such exisited in intervening galaxies as defined here.

The minimum column density in H looking through the disk is 10^{20} cm^{-2}, most of which has less than a factor of 2 depletion in any element (the "halo" gas of Figure 1). The depletion of Fe and Si is, for instance, about 80% in gas with N(H I)$=10^{19}$ cm^{-2} in some clouds within 100 pc that would be considered "intercloud" (York & Kinahan 1979), but over longer pathlengths through the halo, additional clouds have less depletion.. In the disk, 10^{20} H atoms/cm^{-2} would lead to a color excess, E(B-V) of 0.05 (Spitzer 1978). The earlier statements about carbon depletion and the depletion of the heavier elements, together with the low depletion in halo sight lines, suggest that in an intervening galaxy, the *maximum* E(B-V) would be 0.05, corresponding to a maximum total visual absorption of 0.15.

All measurements made so far of low z intervening galaxies in QSOALS show that the total gas phase abundance of Zn, a generally undepleted element in local Milky Way gas, is less than or comparable to 0.1 of the Solar abundance compared to H (Meyer & York 1992; Steidel et al. 1997). All other measured species seem to be low as well. While Fe may show traces of depletion compared to zinc, the amount of material available for depletion onto extinguishing particles is down by 10%, so by the argument given above, there cannot be much extinction. Solar abundances of C and all the other elements are just barely able to explain the extinction per H atom in the Milky Way. If the abundance in all phases of interstellar gas is down by a factor of 0.1, the amount of extinction must be down by a factor of 10, at least (assuming a constant efficiency of particle formation per depleted atom). Therefore, the total visual absorption (A_v) of the intercloud medium in unseen intervening galaxies toward SNe must be less than or equal to 0.02. The occurrence of $A_v > 0.1$ would be extremely rare (occurring when small, translucent clouds are intersected.)

4.3. *Extinction in the Galaxy of the SNe*

Turning then to extinction created local to the SNe I, this too is unlikely to be very important, but can, in principle, be observationally checked, with large telescopes. Even the lowest redshift QSOALS show little differential depletion of elements and abundances are < 0.1 of the solar values (Meyer & York 1992; Steidel et al. 1997), implying little material available to make any kind of grains or large molecules. If the galaxies containing SNe I at high z are L$_*$ galaxies with solar metallicity disks, depletion such as seen in our Galaxy might be expected (in terms of elements depleted into continuous absorbers, per H atom). The accident of seeing SNe of Type I behind heavily absorbing clouds should not happen often, as noted above. In general, the extinction will be the relatively small amount expected from the intercloud medium. This conclusion is consistent with remarks made earlier about the SMC, as a prototype for unseen intervening galaxies.

Little is known about the nature of galaxies at high redshift. We have suggested (York et al. 1986) that galaxies near $z \sim 1$ are agglomerations of dwarf galaxies, in which case the extinction might be like that in the SMC, i.e., minimal. Even if the disks formed at higher redshift, there may not have been time for heavy elements to form in adequate numbers to permit Galactic-like extinction per H atom. In fact, the grains and/or molecules themselves may not have had time to form, in sustainable quantities, even given solar abundances.

4.4. *Observational Tests*

Clearly, some observational test for extinction local to SNeI is needed. The classic test of differential extinction is not valid as that requires a background source with a perfectly known spectrum (which spectrum is what we are trying to correct for by deducing the extinction). If high resolution spectra could be obtained against the SNe themselves, depletion studies similar to those cited above could be used. Such observations would require large telescopes such as are only available on the ground. For objects at redshifts below 2, the lines redshifted into the observable region of the spectrum are not adequate to determine general depletions. There might be some hope of eventually using indirect arguments to infer depletions from lines of Na^o, Ca$^+$, Mg^o, Mg$^+$ and Fe$^+$.

Features identifiable with moderate spectroscopic baselines (a few hundred Angstroms) which have known strengths per unit of differential extinction are called for. The features must be observable in low dispersion spectra of faint objects (the SNe themselves). The features in the extinction curve that may be relevant here are the 2175 Å bump (Mathis 1990; Mathis et al. 1977) and the broadest of the plethora of diffuse interstellar bands in the optical region of the spectrum (Herbig 1995; Jenniskens & Desert 1994). Relationships between the latter bands and extinction have been published (Snow et al. 1977; Wu et al. 1981; Sneden et al. 1978). However, no calibration exists in SMC-like environments. We are undertaking a re-calibration of the relationships that will include environments similar to those in which SNe I at $z \gtrsim 0.5$ might form.

5. Conclusions

More studies are needed to get better statistics on extinction within galaxies at $0.5 < z < 1.5$. The fraction of intervening systems in QSOs that show depletion in this redshift range will be the roughly the same as the fraction of SNe I in that redshift range that show depletion, hence extinction. As one goes to higher and higher z, the corrections needed should be less and less important. Total extinction (grey and selective) cannot be greater, as the abundance of C and all elements is significantly lower. All indications are that even for $z < 0.5$, the effect of intervening systems will be very small.

No disk-like absorption, with solar metalicites and high depletion has been encountered, even in low z QSO absorbers. Statistical studies of low z QSO absorbers can be used to see if any at all show up. Whatever the explanation for the dearth of solar metallicity systems (a combination of the low overall abundance in outlying parts of local groups and the very small cross section of solar metallicty gas?), there is again no evidence that distant QSOs or SNe I will be affected by intervening extinction. Substantial amounts of "snowball" material are unlikely because 1) snowballs have only half or less the extinguishing power of small grains for the same mass; 2) there is little extra carbon to be used for grains aggregated into snowballs when the observed amount of diffential depletion is accounted for, locally; 3) no differential depletion of elements into the cold cloud Galactic pattern is seen at all in most high z cases, so, unless the composition of snowballs exactly mimics solar compostion, elements like oxygen, not involved in differential extinction, are also not appearing in snowballs; and 4) on cosmic scales, the total carbon abundance drops with redshift along with O, Si, Fe, etc., leaving even less room for material locked up in substantial snowballs.

A corollary of the above arguments is that Olbers' was wrong in his explanation of darkness of the night sky.

REFERENCES

AANNESTAD P. & PURCELL E. M., 1973, ARAA, 11, 309

BELL M. B., FELDMAN P. A., TRAVERS M. J., McCARTHY M. C., GOTTLIEB C. A., & THADDEUS P., 1997, ApJ, 483, L61

BROTEN N. W., OKA T., AVERY L. W., MacLEOD J. M. & KROTO H. W., 1978, ApJ, 223, L195

CARDELLI J. A., 1994, Science, 265, 209

CHESAUX J. P. L., 1744, In Traite de la Comete qui a paru en Decembre 1743 & en Janvier, Fevrier & Mars 1744, (Lausanne: Bousquet et Compagni) (Appendix 2, 223-229, translated in reference Harrison 1987)

CROWE M. J., 1994, (New York: Dover), see footnote 2, page 102

CURTIS H. D., 1921, Bulletin of the National Research Council 3, 194, 213

DONN B., 1968, ApJ, 158, L129

DOUGLAS A. E., 1977, Nature, 269, 130

FIELD G. B., 1974, ApJ, 187, 453

FIELD G. B., 1975, in The Dusty Universe, Eds. G. B. Field & A. G. W. Cameron (Neale Watson Academic Publications)

GREENBERG J. M., 1968, in Nebulae and Interstellar Matter, Eds. G. Kuiper & B. M. Middlehurst (University of Chicago Press) 221

GREVESSE N. & ANDERS E., 1998, in Cosmic Abundances of Matter, Ed. C. Jake Waddington (AIP) 1

HARRISON E., 1987, Darkness at Night (Cambridge, MA: Harvard University Press)

HERBIG G. H., 1995, ARAA, 33, 19

HERBIG G., 1993, ApJ, 407, 102

HOBBS L. M., YORK D. G. & OEGERLE W., 1982, ApJ, 257, 135

HOBBS L. M., WELTY D. E., MORTON D. C., SPITZER L. & YORK D. G., 1993, ApJ, 411, 750

JENKINS E. B. & SHAYA E. J., 1979, ApJ, 231, 55

JENNISKENS P. & DESERT F.-X., 1994, A&A, 106, 39

MATHIS J. S., 1990, ARAA, 28, 37

MATHIS J. S., RUMPL W. & NORDSIECK K. H., 1977, ApJ, 217, 425

MCCALL B. J., BEBALLE T. R., HINKLE K. H. & OKA T., 1998, Science, 279, 1910

McCARTHY M. C., TRAVERS M. J., KOVACS A., GOTTLIEB C. A. & THADDEUS P., 1997, ApJS, 113, 105

MEYER D. M., JURA M. & CARDELLI J. A., 1998, ApJ, 493, 222

MEYER D. M. & YORK D. G., 1992, ApJ, 399, L121

OLBERS H. W. M., 1823, in Astronomisches Jahrbuch fur das Jahr 1826 (Berlin: C. R. E. Spathen)

OSTRIKER J. P. & HEISLER J., 1984, ApJ, 278, 1

OSTRIKER J. P., VOGELEY J. & YORK D. G., 1990, ApJ, 364, 540

PEI Y. C., FALL S. M. & BECHTOLD J., 1991, ApJ, 378, 6

PLATT J. R., 1956, ApJ, 123, 486

PURCELL E. M., 1969, ApJ, 158, 433

SALAMA F., BLAKES E. L., ALLAMONDOLA L. J. & TIELENS, A. G. G. M., 1996, ApJ, 458, 621

SMITH W. E., SNOW T. P. & YORK, D. G., 1977, ApJ, 218, 124

SPITZER L. 1978, Physical Processes in the Interstellar Medium (New York: John Wiley and Sons)

SOFIA U. J., CARDELLI J. A., GUERIN K. P. & MEYER D. M., 1997, ApJ, 482, L105

SOFIA U. J., CARDELLI J. A. & SAVAGE B. D., 1994, ApJ, 430, 650

SNEDEN C., GEHRZ R. D., HACKWELL J. A., YORK D. G. & SNOW T. P., 1978, ApJ, 223, 168

SNOW T. P., YORK D. G. & WELTY D. E., 1977, AJ, 82, 113

STEBBINS J., HUFFER C. M. & WHITFORD A. E., 1940, ApJ, 92, 193

STEIDEL C. C., DICKINSON M., MEYER D. M., ADELBERGER K. L. & SEMBACH K. R., 1997, ApJ, 480, 568

TIMMES F. X., LAUROESCH J. T. & TRURAN J. W., 1995, ApJ, 451, 468

TRUMPLER R. J., 1930, Lick Observatory Bulletin, 14, 154

TULEJ M., KIRKWOOD D. A., MACCAFERRI G., DOPFER O. & MAIER J. P., 1998a, Chem. Phys., 228, 293

TULEJ M., KIRKWOOD D. A., PACKLOV M. & MAIER J. P., 1998b, ApJ, 506, L69

VAN DE HULST H. C., 1957, Light Scattering by Small Particles (New York: Wiley and Sons, Inc.)

WELTY D. E., FRISCH P. C., SONNEBORN G. & YORK D. G., 1999, ApJ, in press

WELTY D. E., LAUROESCH J. T., BLADES J. C., HOBBS L. M. & YORK D. G., 1997, ApJ, 489, 672

WU C. C., YORK D. G. & SNOW T. P., 1981, AJ, 86, 755

YORK D. G., DOPITA M., GREEN R. & BECHTOLD J., 1986, ApJ, 311, 610

YORK D. G., & KINAHAN B. F., 1979, ApJ, 228, 127

The Progenitors of Type Ia Supernovae

By MARIO LIVIO

Space Telescope Science Institute, 3700 San Martin Drive, Baltimore, MD 21218 USA

Models for Type Ia Supernovae (SNe Ia) are reviewed. It is shown that there are strong reasons to believe that SNe Ia represent thermonuclear disruptions of C–O white dwarfs, when these white dwarfs reach the Chandrasekhar limit and ignite carbon at their centers.

Progenitor scenarios are reviewed critically and the strengths and weaknesses of each scenario are explicitly presented. It is argued that single-degenerate models, in which the white dwarf accretes from a subgiant or giant companion are currently favored. The relation of the different models to the use of SNe Ia for the determination of cosmological parameters is also discussed.

Observational tests of the conclusions are suggested.

1. Introduction

During the past year two groups have presented strong evidence that the expansion of the universe is accelerating rather than decelerating (Perlmutter et al. 1998, 1999; Schmidt et al. 1998; Riess et al. 1998). This surprising result comes from distance measurements to about fifty supernovae Type Ia in the redshift range $z = 0$ to $z = 1$. The results are consistent with the cosmological constant (or vacuum energy) contributing to the total energy density about 70% of the critical density.

This unexpected finding, as well as the use of supernovae Type Ia to measure the Hubble constant (e.g. Sandage et al. 1996; Saha et al. 1997), has focused the attention again on the frustrating fact that in spite of decades of research the exact nature of the progenitors of supernovae Type Ia remains unknown. Until this problem is solved, one cannot be fully confident that supernovae at higher redshifts are not somehow different from their low redshift counterparts. In the present review I therefore examine critically the question of the nature of the progenitors of supernovae Type Ia. Other recent reviews include Branch et al. (1995), Livio (1996a), Renzini (1996) and Iben (1997).

2. Characteristics and the basic model

The *defining* characteristics of supernovae Type Ia (SNe Ia) are both spectral: (i) the *lack* of lines of hydrogen, and (ii) the *presence* of a strong red Si II absorption feature ($\lambda 6355$ shifted to ~ 6100 Å).

Once defined as SNe Ia, the following are several of the important *observational characteristics* of the class which may help in the search for progenitors:

(1) *Homogeneity*: Nearly 90% of all SNe Ia form a homogeneous class in terms of their *spectra*, *light curves*, and *peak absolute magnitudes*. The latter are given by

$$M_{\rm B} \simeq M_{\rm V} \simeq -19.30(\pm 0.03) + 5\log(H_0/60 \text{ km s}^{-1} \text{ Mpc}^{-1}) \tag{2.1}$$

with a dispersion of $\sigma(M_{\rm B}) \sim \sigma(M_{\rm V}) \sim 0.2$–$0.3$ (Hamuy et al. 1996a; Tamman & Sandage 1995; and see Branch 1998 for a review).

(2) *Inhomogeneity*: Some differences in the spectra and light curves do exist (e.g. Hamuy et al. 1996b). In terms of explosion strength, SNe Ia can roughly be ordered as follows: SN 1991bg and SN 1992K represent the weakest events, followed by weak events like 1986G, followed by about 90% of all SNe Ia which are called "normals" (or "Branch normals"), to the stronger than normal events like SN 1991T.

(3) The *luminosity function* of SNe Ia declines very steeply on the bright side (e.g. Vaughan et al. 1995). Since selection effects cannot prevent the discovery of SNe which are brighter than the "normals," this implies that *the normals are essentially the brightest*.

(4) Near maximum light, the spectra are characterized by *high velocity* (8000–30,000 km s^{-1}) *intermediate mass elements* (O–Ca). In the late, nebular phase, the spectra are dominated by forbidden lines of iron (e.g. Kirshner et al. 1993; Wheeler et al. 1995; Ruiz-Lapuente et al. 1995; Gómez et al. 1996; Filippenko 1997).

(5) Fairly young populations appear to be most efficient at producing SNe Ia (e.g. they tend to be associated with spiral arms in spirals; Della Valle & Livio 1994; Bartunov, Tsvetkov & Filimonova 1994), but relatively old populations ($\tau \gtrsim 4 \times 10^9$ yr) can also produce them. In particular, *SNe Ia do occur in ellipticals* (e.g. Turatto, Cappellaro & Benetti 1994). This immediately implies that *SNe Ia are not caused by the core collapse of stars more massive than 8 $M_\odot$.*

(6) There exist a number of correlations between different pairs of observables (see e.g. Branch 1998 for a review). Of these, the most frequently used in the context of determinations of cosmological parameters is the correlation between the *absolute magnitude and the shape of the light curve*. Basically, brighter SNe Ia decline more slowly. The parameter commonly used to quantify the light curve shape is Δm_{15} (Phillips 1993), the decline in magnitudes in the B band during the first 15 days after maximum light. Hamuy et al. (1996a) find slopes $dM_B/d\Delta m_{15} = 0.78 \pm 0.17$, $dM_V/d\Delta m_{15} = 0.71 \pm 0.14$, and $dM_I/d\Delta m_{15} = 0.58 \pm 0.13$. Using a stretch-factor s (Perlmutter et al. 1997), one can write $M_B = M_B(s=1) - \alpha * (s-1)$, with $M_B(s=1) = -19.46$ (e.g. Sandage et al. 1996), and $\alpha = 1.74$ (Perlmutter et al. 1999). Sophisticated techniques for using the different correlations in distance determinations have been developed (e.g. Riess et al. 1996, 1998).

The above characteristics can be augmented with the following suggestive facts:

(1) The *energy* per unit mass, $1/2(\sim 10^4$ km s$^{-1})^2$, is of the order of the one obtained from the conversion of carbon and oxygen to iron.

(2) The fact that the event is explosive suggests that *degeneracy* may play a role.

(3) The spectrum contains no hydrogen.

(4) The explosions can occur with long delays, after the cessation of star formation.

All the properties above have led to one agreed upon model: *SNe Ia represent thermonuclear disruptions of mass accreting white dwarfs*.

It is interesting that there exists a unanimous consensus on this model in spite of the fact that the essence of flame physics and the details of the transition from deflagration to detonation (in particular the density at which the transition occurs), which are at the heart of the model, remain as major unsolved problems (e.g. Khokhlov, Oran & Wheeler 1997; Woosley 1997; Reinecke, Hillebrandt & Niemeyer 1998; and talks by Khokhlov, Arnett, and Hillebrandt presented at the Chicago meeting on Type Ia Supernovae: Theory and Cosmology, October 1998). In fact, given these uncertainties in the deflagration to detonation transition it is almost difficult to understand how the entire family of SNe Ia light curves can be fitted essentially with one parameter (e.g. Perlmutter et al. 1997), although it is possible that all SNe Ia explode at the same WD mass (see §4), and that the entire observed diversity stems from different ^{56}Ni masses.

3. Why is identifying the progenitors important?

The fact that we do not know yet what are the progenitor systems of some of the most dramatic explosions in the universe has become a major embarrassment and one of the key unsolved problems in stellar evolution. There are several important reasons why identifying the progenitors has become more crucial than ever:

(i) The use of SNe Ia as one of the main ways to determine the key cosmological parameters H_0, and the contribution to the energy density Ω_M, Ω_Λ requires an understanding of the evolution of the luminosity, and the SN rate with cosmic epoch. Both of these depend on the nature of the progenitors.

(ii) Galaxy evolution depends on the radiative, kinetic energy, and nucleosynthetic output of SNe Ia (e.g. Kauffmann, White & Guiderdoni 1993).

(iii) Due to the uncertainties that still exist in the explosion mechanism itself, a knowledge of the initial conditions and of the distribution of matter in the environment of the exploding star are essential for the understanding of the explosion.

(iv) An unambiguous identification of the progenitors, coupled with observationally determined SNe Ia rates will help to place meaningful constraints on the theory of binary star evolution (e.g. Livio 1996b; Li & van den Heuvel 1997; Yungelson & Livio 1998; Hachisu, Kato & Nomoto 1999). In particular, a semi-empirical determination of the elusive common-envelope efficiency parameter, α_{CE}, may be possible (e.g. Iben & Livio 1993).

4. Refinements of the basic model

The basic model for SNe Ia (that essentially all researchers in the field agree upon) is that of a thermonuclear disruption of an accreting white dwarf (WD). However, additional refinements to the model are possible on the basis of existing observational data and theoretical models. These refinements still do not involve the question of the *progenitor systems*. Rather, they address the question of the WD *composition*, and of its *mass* at the instant of explosion.

4.1. *The composition of the exploding WD*

In principle, the WD that accretes to the point of explosion could be composed of He, of C–O, or of O–Ne. Let us examine these possibilities one by one.

(i) *He WDs*: Helium WDs have typical masses that are smaller than ~ 0.45 M$_\odot$ (e.g. Iben & Tutukov 1985). While if accreting, these He WDs can explode following central He ignition at ~ 0.7 M$_\odot$, the composition of the ejected matter in this case will be that of He, ^{56}Ni and decay products (e.g. Nomoto & Sugimoto 1977; Woosley, Taam & Weaver 1986). This is entirely inconsistent with observations (observational characteristic (4) in §2). Therefore, *He WDs certainly do not produce the bulk of SNe Ia*.

(ii) *O–Ne WDs*: Oxygen–Neon WDs form in binaries from main sequence stars of ~ 10 M$_\odot$, although the precise range which allows formation is somewhat uncertain (e.g. Iben & Tutukov 1985; Canal, Isern & Labay 1990; Dominguez, Tornambé & Isern 1993). These systems are probably not numerous enough to constitute the main channel of SNe Ia (e.g. Livio & Truran 1992). It is also generally expected that O–Ne WDs that manage to accrete enough material to reach the Chandrasekhar limit will produce preferentially accretion-induced collapses (to form neutron stars) rather than SNe Ia (e.g. Nomoto & Kondo 1991; Gutierrez et al. 1996). I should note that the existing calculations have been performed for WDs of O–Ne–Mg composition, while recent calculations of the evolution of a 10 M$_\odot$ star produce degenerate cores which are almost devoid of magnesium (Ritossa, Garcia-Berro & Iben 1996). Nevertheless, because of the above two points *it is unlikely that O–Ne WDs produce the bulk of SNe Ia*.

(iii) *C–O WDs*: Carbon–Oxygen WDs are formed in binaries from main sequence stars of up to ~ 10 M$_\odot$. They are therefore both relatively numerous, and they provide a significant "phase space volume" (masses in the range 0.8–1.2 $M_\odot$; accretion rates in the range 10^{-8}–10^{-6} $M_\odot$/yr) in which they are expected to produce SNe Ia (upon reaching the Chandrasekhar limit; e.g. Nomoto & Kondo 1991). Consequently, *the accreting WDs that produce most of the SNe Ia are very probably of C–O composition!*

4.2. *At what mass does the WD explode and where and in what fuel does the ignition take place?*

While there is virtually unanimous agreement about everything I said up to now, namely, that: *SNe Ia are thermonuclear disruptions of accreting C–O WDs*, the next step in the refinement of the model is more controversial. Two major classes of models have been considered, and they suggest entirely different answers to the questions posed by the title of this subsection. In one class, the WD explodes upon reaching the *Chandresekhar mass*, as *carbon* ignites at its *center*. In the second, the WD explodes at a *sub-Chandresekhar mass*, as *helium* ignites *off-center*. I will now review briefly each of these classes and point out its strengths and weaknesses.

4.2.1. *Chandrasekhar mass carbon ignitors*

In this model, considered 'standard,' the WD accretes until it approaches the Chandrasekhar mass. Carbon ignition occurs at or very near the center and the burning front propagates outwards. The main *strengths* of this model are (see e.g. Hoeflich & Khokhlov 1996; Nugent et al. 1997 for detailed modeling):

(1) Some 10^{51} *ergs of kinetic energy* are deposited into the ejecta by nuclear energy.

(2) ^{56}Ni decay powers the *lightcurve*.

(3) The density and composition as a function of the ejection of velocity ($X_i(V_{ej})$) are consistent with the observed *spectra*.

(4) The fact that the explosion occurs at the Chandrasekhar mass explains the *homogeneity*.

(5) Spectra (e.g. of SNe 1994D, 1992A) can be fitted in great detail by theoretical models (e.g. Nugent et al. 1997).

The main *weaknesses* of the Chandrasekhar mass models are:

(1) It has proven more difficult than originally thought for WDs to accrete *up to the Chandrasekhar mass* in sufficient numbers to account for the SNe Ia rate. The difficulty is associated with mass loss episodes in nova explosions, in helium shell flashes and in massive winds or common envelope phases. I will return to some of these problems when I discuss specific progenitor models.

(2) For initial WD masses larger than ~ 1.2 $M_\odot$, *accretion-induced collapse* is a more likely outcome than a SN Ia (e.g. Nomoto & Kondo 1991).

(3) *The late-time spectrum* (~ 300 days), and in particular the Fe III feature at ~ 4700 Å does not agree well with Chandrasekhar mass models (Liu, Jeffrey & Schultz 1998).

My overall assessment of Chandrasekhar mass models is that the strengths significantly overweigh the weaknesses. The calculation of late-time, nebular spectra involves many uncertainties, and hence I do not regard weakness (3) above as fatal (although clearly more work will be required to explain it away). Both weaknesses (1) and (2) can be overcome if it can be demonstrated that SNe Ia statistics can be reproduced within the uncertainties that still plague the theoretical population synthesis models. As I will show in §5, this is indeed the case.

4.2.2. *Sub-Chandrasekhar mass helium ignitors*

In these models a C–O WD accumulates a helium layer of $\sim 0.15\,M_\odot$ with a total mass that is sub-Chandrasekhar. The helium ignites off-center (at the bottom of the layer), resulting in an event known as "Indirect Double Detonation" (IDD) or "Edge Lit Detonation" (ELD). Basically, one detonation propagates outward (through the helium), while an inward propagating pressure wave compresses the C–O core which ignites off-center, followed by an outward detonation (e.g. Livne 1990; Livne & Glasner 1991; Woosley & Weaver 1994; Livne & Arnett 1995; Hoeflich & Khokhlov 1996; and Ruiz-Lapuente, talk presented at the Chicago meeting on Type Ia Supernovae: Theory and Cosmology, October 1998).

The main *strengths* of ELD (sub-Chandrasekhar) models are:

(1) It is easier to achieve the required *statistics*, since less mass needs to be accreted, and the WD does not need to be extremely massive (e.g. Ruiz-Lapuente, Canal & Burkert 1997; Di Stefano et al. 1997; Yungelson & Livio 1998).

(2) The *late-time spectrum* (in particular the Fe III feature at ~ 4700 Å) agrees better with ELD models.

(3) SNe Ia *light curves* can be reproduced adequately by ELD models (although the light curves rise somewhat faster than observed, due to ^{56}Ni heating; Hoeflich et al. 1997).

The main *weaknesses* of ELD models are:

(1) The *spectra* that are produced by ELD models generally do not agree with observations (e.g. of SN 1994D; Nugent et al. 1997). The agreement is somewhat better for the subluminous SNe Ia (e.g. SN 1991bg; Nugent et al. 1997; Ruiz-Lapuente, talk presented at the Chicago meeting on Supernovae, October 1998), but even there it is not very good.

(2) The *highest velocity ejecta have the wrong composition* (^{56}Ni and He; not intermediate mass elements; also no high velocity C; e.g. Livne & Arnett 1995). This is due to the fact that in these models, essentially by necessity, the intermediate mass elements are sandwiched by Ni and He/Ni rich layers.

(3) Since ELD models allow for a range of WD masses, and since more massive WDs produce brighter SNe, one might expect this model to produce a more gradual decline on the bright side of the *luminosity function*, in contradiction to the observed sharp decline (see §2 characteristic (3)).

My overall assessment of the sub-Chandrasekhar mass model is that the weaknesses (and in particular weakness (2) which appears almost inevitable) greatly overweigh the strengths in terms of this being a model for the bulk of SNe Ia. It is still possible that ELDs may correctly represent some subluminous SNe Ia (e.g. Ruiz-Lapuente, Canal, & Burkert 1997). I should note that Pinto (verbal communication at the Chicago meeting on Supernovae) insists that his ELD models manage to overcome all of the above weaknesses and that they are able to produce excellent fits to both light curves and spectra. By the time of the writing of this review, however, I have unfortunately failed to find published results of these models and hence I cannot comment on them.

4.3. *The favored model*

On the basis of the above discussion the basic model can be further refined, and I tentatively conclude that: *SNe Ia represent thermonuclear disruptions of mass accreting C–O white dwarfs, when these white dwarfs reach the Chandrasekhar limit and ignite carbon at their centers!*

5. The two possible scenarios

The next step in the search for the progenitor systems of SNe Ia is even more controversial. Two possible scenarios have been proposed: (i) The *double-degenerate* scenario, in which two CO WDs in a binary system are brought together by the emission of gravitational radiation and coalesce (Webbink 1984; Iben & Tutukov 1984). (ii) The *single-degenerate* scenario, in which a CO WD accretes hydrogen-rich or helium-rich material from a non-degenerate companion (Whelan & Iben 1973; Nomoto 1982).

In the first scenario the progenitor systems are necessarily *binary WD systems* in which the total mass exceeds the Chandrasekhar mass, and which have binary periods shorter than about thirteen hours (to allow merger within a Hubble time).

In the second scenario the progenitors could be systems like: (i) *Recurrent novae* (both of the type in which the WD accretes hydrogen from a giant like T CrB, RS Oph and of the type in which the WD accretes helium rich material from a subgiant like U Sco, V394 CrA, and Nova LMC 1990#2), (ii) *Symbiotic Systems* (in which the WD accretes from a low mass red giant), or (iii) persistent *Supersoft X-ray Sources* (in which the WD accretes at a high rate $\gtrsim 10^{-7}$ $M_\odot$/yr from a subgiant companion).

I will now examine the strengths and weaknesses of each one of these scenarios.

5.1. *The double-degenerate scenario*

There is no question that binary white dwarf systems are an expected outcome of binary star evolution (e.g. Iben & Tutukov 1984; Iben & Livio 1993). Once the lighter WD (which has a larger radius) fills its Roche lobe, it is entirely dissipated within a few orbital periods, to form a massive disk around the primary (e.g. Rasio & Shapiro 1994). The subsequent evolution of the system depends largely on the accretion rate through this disk (e.g. Mochkovitch & Livio 1990; see discussion below).

The main *strengths* of this scenario are the following:

(1) The *absence of hydrogen* in the spectrum is naturally explained in a model which involves the merger of two C–O WDs. In fact, if hydrogen is ever detected in the spectrum of a SN Ia, this would deal a fatal blow to this model. Tentative evidence for circumstellar $H\alpha$ absorption is SN 1990M was presented by Polcaro and Viotti (1991). However, Della Valle, Benetti & Panagia (1996) demonstrated convincingly that the absorption was caused by the parent galaxy, rather than by the SN environment.

(2) In spite of some impressions to the contrary, *many double WD systems do exist*. In a sample of 153 field WDs and subdwarf B stars, Saffer, Livio & Yungelson (1998) found 18 new double-degenerate candidates. There are currently eight known systems with orbital periods of less than half a day. While only one of those systems (KPD 0422+5421; Koen, Orosz & Wade (1998)) has a total mass which within the errors could be higher than the Chandrasekhar mass, the sample of confirmed short-period double-degenerates is still smaller than the number predicted to contain a massive system.

(3) Population synthesis calculations predict the *right statistics* for mergers, about 10^{-3} yr^{-1} events for populations that are $\sim 10^{8}$ yr old and 10^{-4} yr^{-1} for populations that are $\sim 10^{10}$ yr old.

(4) Since double WD systems were found to exist, *mergers* with some "interesting" consequences (either a SN Ia or an accretion-induced collapse) appear inevitable.

(5) The explosion or collapse is expected to occur at the Chrandrasekhar mass, which as I noted in §4.3, I regard as a property of the favored model.

The main *weaknesses* of the double-degenerate scenario are the following:

(1) There are strong indications that WD mergers may lead to off-center carbon ignition, accompanied by the conversion of the C–O WD to an O–Ne–Mg composition, and

followed by an accretion-induced collapse rather than a SN Ia (e.g. Mochkovitch & Livio 1990; Saio & Nomoto 1985, 1998; Woosley & Weaver 1986).

(2) Galactic chemical evolution results, and in particular the behavior of the [O/Fe] ratio as a function of metallicity ([Fe/H]) have been claimed to be inconsistent with WD mergers as the mechanism for SNe Ia (Kobayashi et al. 1998).

Since we are now getting to the final stages in the identification of the progenitors, it is important to assess critically the severity of the above weaknesses. I will therefore discuss now each one of them in some detail.

5.1.1. *Constraints from Galactic chemical evolution*

Supernovae Type II (SNe II) are explosions resulting from the core collapse of massive ($\gtrsim 8$ $M_\odot$) stars. These supernovae produce relatively more oxygen and magnetism than iron ([O/Fe] > 0). On the other hand SNe Ia produce mostly iron and little oxygen. Until recently, the impression has been that metal poor stars ([Fe/H] ≤ -1) have a nearly flat relation of [O/Fe] vs. [Fe/H], with a value of [O/Fe] ~ 0.45 (e.g. Nissen et al. 1994), while disk stars ([Fe/H] $\gtrsim -1$) show a linearly decreasing [O/Fe] with increasing metallicity (e.g. Edvardsson et al. 1993). The break near [Fe/H] ~ -1 was traditionally explained by the fact that the early heavy element production was done exclusively by SNe Ia, with the break occurring when the larger Fe production by SNe Ia kicks in (e.g. Matteucci & Greggio 1986).

Recently, Kobayashi et al. (1998) performed chemical evolution calculations for both the double-degenerate scenario and for the single-degenerate scenario. For the latter they used two types of progenitor systems: one with a red giant companion and an orbital period of tens to hundreds of days, and the other with a near main sequence companion and a period of a few tenths of a day to a few days.

They obtained for the double-degenerate scenario (for which they took a time delay of ~ 0.1–0.3 Gyr) a break at [Fe/H] ~ -2. For the single-degenerate scenario (with a delay caused by the main sequence lifetime of $\gtrsim 1$ Gyr; including metallicity effects), they obtained a break at [Fe/H] ~ -1. Kobayoshi et al. (1998) thus concluded that the double-degenerate scenario is inconsistent with Galactic chemical evolution results.

Personally, I am not convinced by this apparent discrepancy, since Galactic chemical evolution calculations (and observations) are notoriously uncertain. In particular, the most recent Keck observations of oxygen in unevolved metal-poor stars show *no break* in the [O/Fe] vs. [Fe/H] relation. Rather, oxygen is enhanced relative to iron over three orders of magnitude in [Fe/H] in a robustly linear relation (Boesgaard et al. 1999). Consequently, apparent inconsistencies based on Galactic chemical evolution cannot be regarded at present as a fatal weakness of the double-degenerate scenario.

5.1.2. *SN Ia or accretion induced collapse?*

Potentially a more serious (and possibly even fatal) weakness of the double-degenerate scenario comes from the fact that some estimates and calculations indicate that the coalescence of two C–O WDs may lead to an accretion-induced collapse rather than to a SN explosion (e.g. Mochkovitch & Livio 1990; Saio & Nomoto 1985, 1998; Kawai, Saio & Nomoto 1987; Timmes, Woosley & Taam 1994).

The point is the following: once the lighter WD fills its Roche lobe, it is dissipated within a few orbital periods (Benz et al. 1990; Rasio & Shapiro 1995; Guerro 1994) and it forms a hot thick disk configuration around the more massive white dwarf. This disk is mainly rotationally supported and hence central carbon ignition does not take place immediately, but rather the subsequent evolution depends largely on the rate of angular momentum transport and removal, since they determine the accretion rate onto the primary WD. As long as the accretion rate is higher than about $\dot{M} \gtrsim 2.7 \times 10^{-6}$ M$_\odot$ yr^{-1}, *carbon is ignited off-center* (at the core-disk boundary; this may happen during the merger itself; e.g. Segretain 1994). Under such conditions, the flame was found (in spherically symmetric calculations) to propagate all the way to the center within a few thousand years, thus burning the C–O into an O–Ne–Mg mixture with *no explosion* (i.e. before carbon is centrally ignited; e.g. Saio & Nomoto 1998). Such configurations are expected to collapse (following electron captures on ^{24}Mg) to form neutron stars (Nomoto & Kondo 1991; Canal 1997). The main questions are then:

(i) What accretion rates can be expected from the initial WD-thick disk configuration?

(ii) May some aspects of the flame propagation be different given the fact that the real problem is three-dimensional while most of the existing calculations were performed using a spherically symmetric code? In particular, *could the carbon burning be quenched before the transformation to O–Ne–Mg composition occurs?*

The answers to both of these questions involve uncertainties, however some possibilities are more likely than others. First, it appears *very difficult to avoid high accretion rates*. If the MHD turbulence that is expected to develop in accretion disks (e.g. Balbus & Hawley 1998) is operative, with a corresponding viscosity parameter of $\alpha \sim 0.01$ (where the viscosity is given by $\nu \sim \alpha c_s H$, with H being a vertical scaleheight in the disk and c_s the speed of sound; e.g. Balbus, Hawley & Stone 1996), then angular momentum can be removed in a matter of days! In such a case, even if the accretion rate is Eddington limited (at $\sim 10^{-5}$ M$_\odot$/yr), off-center carbon ignition should still occur, with an eventual collapse rather than an explosion. Deviations from spherical symmetry can only hurt, since they may allow accretion to proceed at a super-Eddington rate. It is difficult to see why the dynamo-generated viscosity would be suppressed for the kind of shear and temperatures expected in the disk.

Concerning the burning itself, recent attempts at multi-dimensional calculations of the flame propagation and a more detailed analysis of some of the processes involved (Garcia-Senz, Bravo & Serichol 1998; Bravo & Garcia-Senz 1999) indicate that if anything, accretion induced collapses are an even more likely outcome than previously thought. This is due to the effects of electron captures in Nuclear Statistical Equilibrium which tend to stabilize the thermonuclear flame, and to Coulomb corrections to the equation of state. The latter has the effect of reducing the flame velocities and the electronic and ionic pressures, all of which result in a reduction in the critical density which separates explosions from collapses.

Finally, on the observational side there are also two points which argue at some level against WD mergers as SNe Ia progenitors.

(i) Even if MHD viscosity could somehow be suppressed, and the disk surrounding the primary WD could cool down, so that angular momentum would be transported only via the viscosity of (partially) degenerate electrons, this would result in an accretion timescale of $\sim 10^9$ yrs (Mochkovitch & Livio 1990; Mochkovitch et al. 1997). The system prior to the explosion would have an absolute magnitude of $M_V \lesssim 10$ (with much of the emission occurring in the UV). There is no evidence for the existence of some $\sim 10^7$ such objects in the Galaxy.

(ii) The existence of planets around the pulsars PSR 1257+12 and PSR 1620–26 (Wol-szczan 1997; Backer 1993; Thorsett, Arzoumanian & Taylor 1993) could be taken to mean (this is a model dependent statement) that mergers tend to produce accretion induced collapses rather than SNe Ia. In one of the leading models for the formation of such planets (Podsiadlowski; Pringle & Rees 1991; Livio, Pringle & Saffer 1992), the planets form in the following sequence of events. The lighter WD is dissipated (upon Roche lobe overflow) to form a disk around the primary. As material from this disk is accreted, matter at the outer edge of the disk has to absorb the angular momentum, thereby expanding the disk to a large radius. The planets form from this disk in the some way that they did in the solar system, while the central object collapses to form a neutron star.

5.1.3. *Overall assessment of the double-degenerate scenario*

It has now been observationally demonstrated that many double-degenerate systems exist. The general agreement between the distribution of the observed properties (e.g. orbital periods, masses) and those predicted by population synthesis calculations (Saffer, Livio & Yungelson 1998), suggests that the fact that no clear candidate (short period) system with a total mass exceeding the Chandrasekhar mass has been found yet, may merely reflect the insufficient size of the observational sample. Thus, there is very little doubt in my mind that statistics is not a serious problem. The most disturbing uncertainty is related to the outcome of the merger process itself. The discussion in §5.1.2 suggests that *collapse to a neutron star is more likely than a SN Ia* (see also Mochovitch et al. 1997).

5.2. *The single-degenerate scenario*

The main *strengths* of the single degenerate scenario are:

(1) A class of objects in which hydrogen is being transferred at such high rates that it *burns steadily* on the surface of the WD has been identified—the Supersoft X-ray Sources (Greiner, Hasinger & Kahabka 1991; van den Heuvel et al. 1992; Southwell et al. 1996; Kahabka and van den Heuvel 1997). If the accreted matter can indeed be retained, this provides a natural path to an increase in the WD mass towards the Chandrasekhar mass (e.g. Di Stefano & Rappaport 1994; Livio 1995, 1996a; Yungelson et al. 1996).

(2) Other candidate progenitor systems are known to exist, like symbiotic systems (e.g. Munari & Renzini 1992; Kenyon et al. 1993; Hachisu, Kato & Nomoto 1999) and recurrent novae (Hachisu et al. 1999).

(3) There have been claims that the single degenerate scenario fits better the results of Galactic chemical evolution (e.g. Kobayoski et al. 1998). However, as I have shown in §5.1.1, recent observations cast doubt on this assertion. Similarly, nucleosynthesis results show that in order to avoid unacceptably large ratios of $^{54}Cr/^{56}Fe$ and $^{50}Ti/^{56}Fe$, the central density of the WD at the moment of thermonuclear runaway must be lower than $\sim 2 \times 10^9$ g cm^{-3} (Nomoto et al. 1997). Such low densities are realized for high accretion rates ($\gtrsim 10^{-7}$ M$_\odot$ yr^{-1}), which are typical for the Supersoft X-ray Sources. Nucleosynthesis results suffer too, however, from considerable uncertainties (e.g. Nagataki, Hashimoto & Sato 1998).

The main *weaknesses* of the single degenerate scenario are:

(1) The upper limits on *radio detection* of hydrogen at 2 and 6 cm in SN 1986G, taken approximately one week before optical maximum (Eck et al. 1995), rule out a symbiotic system progenitor for this system with a wind mass loss rate of $10^{-7} \lesssim \dot{M}_W \lesssim 10^{-6}$ M$_\odot$ yr^{-1} (Boffi & Branch 1995). This in itself is not fatal, since SN 1986G is somewhat peculiar (e.g. Branch and van den Bergh 1993), and the upper limit on the mass loss rate is at the high end of observed symbiotic winds. An even less stringent upper limit from x-ray and Hα observations exists for SN 1994D (Cumming et al. 1996).

(2) There exists some uncertainty whether WDs can then reach the Chandrasekhar mass *at all* by the accretion of hydrogen (e.g. Cassisi, Iben & Tornambe 1998). Furthermore, even if they can, the question of whether they can produce the required SNe Ia statistics is highly controversial (e.g. Yungelson et al. 1995, 1996; Yungelson & Livio 1998; Hachisu, Kato & Nomoto 1999; Hachisu et al. 1999).

I will now examine these weaknesses in some detail.

5.2.1. *Observational detection of hydrogen*

Ultimately, the presence or total absence of hydrogen in SNe Ia will distinguish unambiguously between single-degenerate and double-degenerate models. To date, hydrogen has not been convincingly detected in *any* SN Ia. It is interesting to note that narrow $\lambda 6300$, $\lambda 6363$ [OI] lines were observed only in one SN Ia (SN 1937C; Minkowski 1939), but even in that case there was no hint of a narrow Hα line. Hachisu, Kato & Nomoto (1999) estimate in one of their models (which involves stripping of material from the red giant; see below) a density measure of $\dot{M}/v_{10} \sim 10^{-8}$ M$_\odot$ yr^{-1} (where v_{10} is the wind velocity in units of 10 km s^{-1}), while the most stringent radio upper limit existing currently (for SN 1986G) is $\dot{M}/v_{10} \sim 10^{-7}$ M$_\odot$ yr^{-1} (Eck et al. 1995; for SN 1994D Cumming et al. (1996) find from Hα an upper limit of $\dot{M} \sim 1.5 \times 10^{-5}$ M$_\odot$ yr^{-1} for a wind speed of 10 km s^{-1}; for SN 1992A Schlegel & Petre (1993) find from X-ray observations an upper limit of $\dot{M}/v_{10} = (2-3) \times 10^{-6}$ M$_\odot$ yr^{-1}). Thus, while it is impossible at present to rule out single-degenerate models on the basis of the apparent absence of hydrogen, the hope is that near future observations will be able to determine definitively whether this absence is real or if it merely represents the limitations of existing observations (an improvement by two orders of magnitude will give a definitive answer).

5.2.2. *Statistics*

Growing the WD to the Chandrasekhar mass is not easy. At accretion rates below $\sim 10^{-8}$ M$_\odot$/yr WDs undergo repeated nova outbursts (e.g. Prialnik & Kovetz 1995), in which the WDs lose more mass than they accrete between outbursts (e.g. Livio & Truran 1992). For accretion rates in the range 10^{-8}–a few $\times 10^{-7}$ M$_\odot$/yr, while helium can accumulate, the WDs experience mass loss due to helium shell flashes and due to the common envelope phase which results from the engulfing of the secondary star in the expanding envelope (with mass loss occurring due to drag energy deposition). At accretion rates above a few $\times 10^{-7}$ M$_\odot$/yr, the WDs expand to red giant configurations and lose mass due to drag in the common envelope and due to winds (e.g. Cassisi et al. 1998). The net result of this has been that population synthesis calculations which follow the evolution of all the binary systems in the Galaxy, tended until recently to conclude that single degenerate channels manage to bring WDs to the Chandrasekhar mass only at about 10% of the inferred SNe Ia frequency of 4×10^{-3} yr^{-1} (e.g. Yungelson et al. 1995, 1996; Yungelson & Livio 1998; Di Stefano et al. 1997; although see Li & van den Heuvel 1997).

Very recently, a few serious attempts have been made to investigate whether the statistics could be improved by increasing the "phase space" for single degenerate scenarios, given the fact that population synthesis calculations involve many assumptions. These attempts resulted in three directions in which the phase space could be increased.

(i) The accumulation efficiency of helium has been recalculated using OPAL opacities (Kato & Hachisu 1999). These authors concluded that helium can accumulate much more efficiently than found by Cassisi et al. (1998), mainly because the latter authors used relatively low WD masses (0.516 M$_\odot$ and 0.8 M$_\odot$) and old opacities in their calculations.

(ii) Hachisu et al. (1999) claimed to have identified an evolutionary channel for single-degenerate systems previously overlooked in population synthesis calculations. In this channel, the C–O WD is formed from a red giant with a helium core of 0.8–2.0 M$_\odot$ (rather than from an asymptotic giant branch star with a C–O core). The immediate progenitors in this case are expected to be either helium-rich Supersoft X-ray Sources or recurrent novae of the U Sco subclass.

(iii) It has been suggested that the inclusion of a few additional physical effects, can increase substantially the phase space of the symbiotic channel (Hashisu, Kato & Nomoto 1999). These new effects included:

(1) The WD loses much of the transferred mass in a massive wind. This has the effect that the mass transfer process is stabilized for a wider range of mass ratios, up to $q_{max} \equiv m_2/m_1 = 1.15$ instead of $q_{max} = 0.79$ without the massive wind.

(2) It has been suggested that the wind from the WD strips the outer layers of the red giant at a high rate. This increases the allowed mass ratios (for stability) even above 1.15, essentially indefinitely.

(3) It has been suggested that at large separations (up to $\sim 30,000\,R_\odot$) the wind from the red giant acts like a common envelope to reduce the separation, thus allowing much wider initial separations to result in interaction.

There are many uncertainties associated with all of these attempts to increase the phase space. For example, the efficiency of mass stripping from the giant by the wind from the WD may be much smaller than assumed by Hachisu et al. (1999), for the following reasons. At high accretion rates, much of the mass loss from the WD may be in the form of an outflow or a collimated jet, perpendicular to the accretion disk rather than in the direction of the giant. Evidence that this is the case is provided by the jet satellite lines to He II 4686, Hβ and Hα observed in the Supersoft X-ray Source RX J0513.9−6951 (Southwell et al. 1996). These jet lines are very similar to those seen in the prototypical jet source SS 433 (e.g. Vermeulen et al. 1992). Furthermore, even if some of the WD wind hits the surface of the giant, it is not clear how efficient it would be in stripping mass, since the rate of energy deposition per unit area by the wind is smaller by two orders of magnitude that the giant's own intrinsic flux.

Similarly, the efficiency of helium accumulation is still highly uncertain, as the differences between the results of Kato & Hachisu (1999) and Cassisi et al. (1998) have shown.

Finally, all the new suggestions for the increase in phase space rely very heavily on the results of the wind solutions of Kato (1990; 1991), which involve a treatment of the radiation not nearly as sophisticated as that of more state of the art radiative transfer codes (e.g. Hauschildt et al. 1995, 1996).

5.2.3. *Overall assessment of the single-degenerate scenario*

The above discussion suggests that probably not all the scenarios for increasing the "phase space" of the single-degenerate channels work (if they did we might have had the opposite problem of too high a frequency of SNe Ia!). However, these attempts serve to demonstrate that the input physics to population synthesis codes still involves many uncertainties. My feeling is therefore that given the many potential channels leading to SNe Ia, statistics should not be regarded as a serious problem.

Single-degenerate scenarios therefore appear quite promising, since unlike the situation a decade ago, a class of objects in which the WDs accrete hydrogen steadily (the Supersoft X-Ray Sources) has actually been identified. The main problem with single-degenerate scenarios remains the non-detection of hydrogen so far. While a difficult observational problem (see §6), the establishment of the presence or absence of hydrogen in SNe Ia should become a first priority for SNe observers.

5.3. *What if....?*

Given the fact that there are still uncertainties involved in identifying the SNe Ia progenitors, and that WD mergers and some form of off-center helium ignitions almost certainly occur, it is instructive to pose a few "what if" questions. For example: *What if WD mergers with a total mass exceeding Chandrasekhar do not produce SNe Ia, what do they produce then?* The answer in this case will have to be that they almost certainly produce either neutron stars via accretion induced collapses, or single WDs, if the merger is accompanied by extensive mass loss from the system.

What if off-center helium ignitions do not produce SNe Ia? In this case, if an explosive event indeed ensues, a population of "super novae" (with ~ 0.15 $M_\odot$ of ^{56}Ni and He) is yet to be detected (maybe SN 1885A in M31 was such an event?). *What if off-center helium ignitions do produce SNe Ia? What comes out of the systems with $M_{WD} \gtrsim 1$ $M_\odot$, which should be even brighter?* It is difficult to believe that the latter are represented by the very few bright objects like SN 1991T. Thus, we see that off-center helium ignitions seem to present an observational problem both if they *do* and if they *do not* produce SNe Ia. To me this suggests that the physics of these events is not well understood (for example, maybe off-center helium ignition fails to ignite the C–O core after all).

6. How can we hope to identify the progenitors?

There are several ways in which observations of *nearby* supernovae could solve the mystery of SNe Ia progenitors:

(1) A combination of *early high resolution optical spectroscopy, x-ray observations* and *radio observations* can both provide limits on $\dot{M}/v$ from the progenitors and potentially detect the presence of circumstellar hydrogen (if it exists).

For example, narrow HI in emission or absorption could be detected either very early, or shortly after the ejecta become optically thin (~ 100 days). The latter is true because the SN ejecta probably engulfs the companion at early times (e.g. Chugai 1986; Livne, Tuchman & Wheeler 1992). The interaction of the ejecta with the circumstellar medium can be observed either in the radio (e.g. Boffi & Branch 1995) or in x-rays (e.g. Schlegel 1995). The collision of the ejecta (with circumstellar matter) can also set up a forward and a reverse shock (e.g. Chevalier 1984; Fransson, Lundqvist & Chevalier 1996), and radiation from the latter can ionize the wind and produce Hα emission (e.g. Cumming et al. 1996).

(2) Early observations of the gamma-ray light curve (or gamma-ray line profiles) could distinguish between carbon ignitors and sub-Chandrasekhar helium ignitor models (see §4.2.2) since the latter can be expected to result in a quicker rise of the gamma-ray light curve due to the presence of ^{56}Ni in the outer layers (and different gamma-ray line profiles; because of the high velocity ^{56}Ni).

Observations of very distant supernovae (at $z \sim 3$–4) with the Next Generation Space Telescope (NGST) can also help (e.g. Yungelson & Livio 1999). For example, the progenitors can be identified from the observed frequency of SNe Ia as a function of redshift (e.g. Yungelson & Livio 1998, 1999; Ruiz-Lapuente & Canal 1998; Madau, Della Valle & Panagia 1998; and see §8), since different progenitor models produce different redshift distributions. Personally, I think that it would be absolutely pathetic to have to resort to this possibility. Rather, one would like to identify the progenitors independently, and to use the observations of supernovae at high z to constrain models of cosmic evolution of rates, luminosity, and input into galaxies.

7. Could we be fooled?

One of the key questions that result from the fact we do not know with certainty which systems are the progenitors of SNe Ia is clearly: *is it possible that SNe Ia at higher redshifts are systematically dimmer than their low-redshift counterparts?* In this respect it is important to remember that a systematic decrease in the brightness by ~ 0.25 magnitudes is sufficient to explain away the need for a cosmological constant. In a recent work, Yungelson & Livio (1999) calculated the expected ratio of the rate of SNe Ia to SNe from massive stars (Types II, Ia, Ic) as a function of redshift for several progenitor models. They showed that if different progenitor systems can contribute to the total SNe Ia rate (e.g. double-degenerates at the Chandresekhar mass and single-degenerates with subgiant donors at sub-Chandrasekhar masses), then it is possible in principle that a different progenitor class will start to dominate at $z \sim 1$. However, such a transition is highly unlikely, because: (i) in some models the transition has the opposite effect to the observed one (e.g. double degenerates which may be expected to be brighter than sub-Chandrasekhar ELDs dominate at the higher redshifts). (ii) If the contribution from physically different channels was indeed significant, one would have expected to observe this division more clearly also in the local sample, which is not the case ($\sim 90\%$ of SNe Ia are "normals"). Consequently, I do not believe that the observed universal acceleration is an artifact of the observed sample being dominated by different progenitor classes.

8. Tentative conclusions and observational tests

On the basis of the analysis and discussion in the present work, the following tentative conclusions can be drawn:

(1) SNe Ia are almost certainly thermonuclear disruptions of mass accreting *C–O white dwarfs*.

(2) It is very likely that the explosion occurs *at the Chandrasekhar mass*, as *carbon is ignited at the WD center*. Off-center ignition of helium at sub-Chandrasekhar masses may still be responsible for a subset of the SNe Ia which are subluminous, but this is less clear.

(3) The immediate progenitor systems are still not known with certainty. From the discussion in §5 (see in particular §5.1.3 and 5.2.3) however, I conclude that presently *single degenerate scenarios look more promising*, with hydrogen or helium rich material being transferred from a subgiant or giant companion (systems like Supersoft X-Ray Sources and Symbiotics).

(4) Definitive answers concerning the nature of the progenitors can be obtained from observations taken as early as possible in: *x-rays, radio, and high resolution optical spectroscopy. The establishment of the presence or absence of hydrogen in SNe Ia should be regarded as an extremely high priority goal for supernovae observers.* If hydrogen will not be detected at interesting limits (corresponding to $\dot{M}/v_{10} \sim 10^{-8}$ M$_\odot$ yr^{-1}), this will point clearly towards the double-degenerate scenario.

(5) Observations of SNe Ia at high redshifts can help to test particular ingredients of the models which are directly related to the nature of the progenitors. For example, most of the models aiming at improving the statistics of the single-degenerate scenarios rely on a strong wind from the accreting WD. These models thus predict an "inhibition" of SNe Ia in low-metallicity environments, and in particular a significant decrease in the rate of SNe Ia at $z \sim 1$–2 (Kobayashi et al. 1998). At present, the detection of a very likely SN Ia at redshift $z = 1.32$ (SN 1997ff) in the Hubble Deep Field appears inconsistent with this prediction (Gilliland, Nugent & Philips 1999), but more observations will be required to give a more definitive answer.

This research has been supported in part by NASA Grant NAG5-6857.

REFERENCES

BACKER D. C. 1993 In ASP 36: Planets Around Pulsars Eds. J. A. Phillips et al. (ASP) 11

BALBUS S. A. & HAWLEY J. F., 1998, Rev. Mod. Phys., 70, 1

BALBUS S. A., HAWLEY J. F. & STONE J. M., 1996, ApJ, 467, 76

BARTUNOV O. S., TSVETKOV D. YU & FILIMONOVA I. V., 1994 PASP, 106, 1276

BENZ W., CAMERON A. G. W., BOWERS R. L., PRESS W. H., 1990, ApJ, 348, 647

BOESGAARD A. M., KING J. R., DELIGANNIS C. P. & VOGT S. S., 1999, AJ, in press.

BOFFI F. R. & BRANCH D. 1995, PASP, 107, 347

BRANCH D., 1998, ARAA, 36, 17

BRANCH D. & VAN DEN BERGH S., 1993, AJ, 105, 2231

BRAVO E. & GARCIA-SENZ D., 1999, MNRAS, submitted

CANAL R., 1997, In Thermonuclear Supernovae, Eds. P. Ruiz-Lapuente, R. Canal & J. Isern, (Kluwer) 257

CANAL R., ISERN J. & LABAY J., 1990, ARAA, 28, 183

CASSISI S., IBEN I. JR. & TORNANBE A., 1998, ApJ, 496, 376

CHEVALIER R. A., 1984, ApJ, 285, L63

CHUGAI N. N., 1986, SvA, 30, 563

CUMMING R. J., LUNDQVIST P., SMITH L. J., PETTINI M. & KING D. L., 1996, MNRAS, 283, 1355

DELLA VALLE M., BENETTI S. & PANAGIA N., 1996, ApJ, 459, 23

DELLA VALLE M. & LIVIO M., 1994, ApJ, 423, L31

DI STEFANO R., NELSON L. A., LEE W., WOOD T. H. & RAPPAPORT S., 1997, In Thermonuclear Supernovae, Eds. R. Ruiz-Lapuente, R. Canal, & J. Isern, (Kluwer) 147

DI STEFANO R. & RAPPAPORT S., 1994, ApJ, 437, 733

DOMINGUEZ I., TORNAMBÉ A. & ISERN J., 1993, ApJ, 419, 268

ECK C., COWAN J. J., ROBERTS D., BOFFI F. R. & BRANCH D., 1995, ApJ, 451, L53

EDVARDSSON B., ANDERSEN J., GUSTOFSSON B., LAMBERT D. L., NISSEN P. E. & TOMKIN J., 1993, A&A, 275, 101

FILIPPENKO A. V., 1997, ARAA, 35, 309

FRANSSON C., LUNDQVIST P. & CHEVALIER R. A., 1996, ApJ, 461, 993

GARCIA-SENZ D., BRAVO E., & SERICHOL N., 1998, ApJS, 115, 119

GILLILAND R. I., NUGENT P. E. & PHILLIPS M. M., 1999, ApJ, in press

GÓMEZ G., LÓPEZ R. & SÁNCHEZ F., 1996, AJ, 112, 2094

GREINER J., HASINGER G. & KAHABKA P., 1991, A&A, 246, L17

GUTIÉRREZ J., GARCIA-BERRO E., IBEN I. JR., ISERN J., LABAY J. & CANAL R., 1996, ApJ, 459, 701

HACHISU I., KATO M. & NOMOTO K., 1999, ApJ, in press (astro-ph/9902304)

HACHISU I., KATO M., NOMOTO K. & UMEDA H., 1999, ApJ, in press (astro-ph/9902303)

HAMUY M., PHILLIPS M. M., SUNTZEFF N. B., SCHOMMER R. A. MAZA J. & AVILÉS R., 1996a, AJ, 112, 2398

HAMUY M., PHILLIPS M. M., SUNTZEFF N. B., SCHOMMER R. A., MAZA J., ET AL., 1996b, AJ, 112, 2438

HAUSCHILDT P. H., BARON E., STARRFIELD S. & ALLARD F., 1996, ApJ, 462, 386

HAUSCHILDT P. H., STARRFIELD S., SHORE S. N., ALLARD F. & BARON E., 1995, ApJ, 447, 829

HOEFLICH P. & KHOKHLOV A., 1996, ApJ, 457, 500

IBEN I. JR. & LIVIO M., 1993, PASP, 105, 1373

IBEN I. JR. & TUTUKOV A. V., 1984, ApJS, 54, 355

IBEN I. JR. & TUTUKOV A. V., 1985, ApJS, 58, 661

KAHABKA P. & VAN DEN HEUVEL E. P. J., 1997, ARAA, 35, 69

KATO M., 1990, ApJ, 355, 277

KATO M., 1991, ApJ, 369, 471

KATO M. & HACHISU I., 1999, ApJ, submitted (astro-ph/991080)

KAUFFMANN G., WHITE S. D. M. & GUIDERDONI B., 1993, MNRAS, 264, 201

KAWAI Y., SAIO H. & NOMOTO K., 1987, ApJ, 315, 229

KENYON S. J., LIVIO M., MIKOLAJEWSKI J. & TOUT C. A., 1993, ApJ, 407, L81

KIRSHNER R. P., et al., 1993, ApJ, 415, 589

KHOKHLOV A. M., ORAN E. S. & WHEELER I. C., 1997, ApJ, 478, 678

KOBAYASHI C., TSUJIMATO T., NOMOTO K., HACHISU I. & KATO M., 1998, ApJ, 503, L155

KOEN C., OROSZ J. A. & WADE R. A., 1998, MNRAS, 300, 695

LI X.-D. & VAN DEN HEUVEL E. P. J., 1997, A&A, 322, L9

LIU W., JEFFREY D. D. & SCHULTZ D. R., 1998, ApJ, 494, 812

LIVIO M., 1995, In Millisecond Pulsars: A Decade of Surprise, Eds. A. S. Fruchter, M. Tavani & D. R. Backer (ASP) 105

LIVIO M., 1996a, In Supersoft X-Ray Sources, Ed. G. Greiner (Springer) 183

LIVIO M., 1996b, In Evolutionary Processes in Binary Stars, Eds. R. A. M. J. Wijers, M. B. Davies & C. A. Tout (Kluwer) 141

LIVIO M., PRINGLE J. E. & SAFFER R. A., 1992, MNRAS, 257, 15p

LIVIO M. & TRURAN J. W., 1992, ApJ, 389, 695

LIVNE E., 1990, ApJ, 354, L53

LIVNE E. & ARNETT D., 1995, ApJ, 454, 62

LIVNE E. & GLASNER A., 1991, ApJ, 370, 272

LIVNE E., TUCHMAN Y. & WHEELER J. C., 1992, ApJ, 399, 665

MADAU P., DELLA VALLE M., & PANAGIA N., 1998, MNRAS, 297, L17

MATTEUCCI F. & GREGGIO L., 1986, A&A, 154, 279

MINKOWSKI R., 1939, ApJ, 89, 156

MOCHKOVITCH R., GUERRERO J. & SEGRETAIN L., 1997, In Thermonuclear Supernovae Eds. P. Ruiz-Laguente, R. Canal & J. Isern (Kluwer) 187

MOCHKOVITCH R. & LIVIO M., 1990, A&A, 236, 378

MUNARI V. & RENZINI A., 1992, ApJ, 397, L87

NAGATAKI S., HASHIMOTO M. & SATO K., 1998, PSAJ, 50, 75

NISSEN P. E., GUSTAFSSON G., EDVARDSSON B. & GILMORE G., 1994, A&A, 285, 440

NOMOTO K., 1982, ApJ, 253, 798

NOMOTO K., IWAMOTO K., NAKASATO N., THIELEMANN F.-K., BRACHWITZ F., TSUJIMOTO T., KUBO Y. & KISHIMOTO N., 1997, Nuc. Phys., A621, 467c

NOMOTO K. & KONDO Y., 1991, ApJ, 367, L19

NOMOTO K. & SUGIMOTO D., 1977, PASJ, 29, 765

NUGENT P., BARON E., BRANCH D., FISHER A. & HAUSCHILDT P. H., 1997, ApJ, 485, 812

PERLMUTTER S. et al., 1997, ApJ, 483, 565

PERLMUTTER S. et al., 1998, astro-ph/9812473

PERLMUTTER S. et al., 1999, ApJ, in press (astro-ph/9812133)

PODSIADLOWSKI PH., PRINGLE J. E. & REES M. J., 1991, Nature, 352, 783

POLCARO V. F. & VIOTTI R., 1991, A&A, 242, L9

PRIALNIK D. & KOVETZ A., 1995, ApJ, 445, 789

RASIO F. A. & SHAPIRO S. L., 1995, ApJ, 438, 887

REINECKE M., HILLEBRANDT W. & NIEMEYER J. C., 1998, A&A, in press (astro-ph/9812120)

RENZINI A., 1996, In Supernovae and Supernova Remnants, Eds. R. McCray & Z. Wang (Cambridge) 23

RIESS A. G., NUGENT P., FILIPPENKO A. V., KIRSHNER R. P. & PERLMUTTER S., 1998, ApJ, 504, 935

RIESS A. G. et al., 1998, AJ, 116, 1009

RIESS A. G., PRESS W. H. & KIRSHNER R. P., 1996, ApJ, 473, 88

RITOSSA C., GARCIA-BERRO E. & IBEN I. JR., 1996, ApJ, 460, 489

RUIZ-LAPUENTE P. & CANAL R., 1998, ApJ, 497, L57

RUIZ-LAPUENTE P., CANAL R. & BURKERT A., 1997. In Thermonuclear Supernovae, Eds. R. Ruiz-Lapuente, R. Canal, & J. Isern (Kluwer) 205

RUIZ-LAPUENTE P., KIRSHNER R. P., PHILLIPS M. M., CHALLIS P. M., SCHMIDT B. P., FILIPPENKO A. V. & WHEELER J. C., 1995, ApJ, 439, 60

SAFFER R. A., LIVIO M. & YUNGELSON L. R., 1998, ApJ, 502, 394

SAHA A., SANDAGE A., LABHARDT L., TAMMANN G. A., MACCHETTO F. D. & PANAGIA N., 1997, ApJ, 486, 1

SAIO H. & NOMOTO K., 1985, A&A, 150, L21

SAIO H. & NOMOTO K., 1998, ApJ, 500, 388

SANDAGE A., SAHA A., TAMMANN G. A., LABHARDT L., PANAGIA N. & MACCHETTO F. D., 1996, ApJ, 460, L15

SCHLEGEL E. M., 1995, Rep. Prog. Phys., 58, 1375

SCHMIDT B. et al., 1998, ApJ, 507, 46

SOUTHWELL K. A., LIVIO M., CHARLES P. A., O'DONOGHUE D. & SUTHERLAND W. J., 1996, ApJ, 470, 1065

TAMMAN G. A. & SANDAGE A., 1995, ApJ, 452, 16

THORSETT S. E., ARZOWMANIAN Z. & TAYLOR J. H., 1993, ApJ, 412, L33

TIMMES F. X., WOOSLEY S. E. & TAAM R. E., 1994, ApJ, 420, 348

TURATTO M., CAPPELLARO E. & BENETTI S., 1994, AJ, 108, 202

VAUGHAN T. E., BRANCH D., MILLER D. L. & PERLMUTTER S., 1995, ApJ, 439, 558

WEBBINK R. F., 1984, ApJ, 227, 355

WHEELER J. C., HARKNESS R. P., KHOKHLOV A. V. & HOEFLICH P., 1995, Phys. Rep., 256, 211

WHELAN J. & IBEN I. JR., 1973, ApJ, 186, 1007

WOLSZCZAN A., 1997, Cel. Mech. & Dyn. Aston., 68, 13

WOOSLEY S. E., 1997, In Thermonuclear Supernovae, Eds. P. Ruiz-Lapuente, R. Canal & J. Isern (Kluwer) 313

WOOSLEY S. E., TAAM R. E. & WEAVER T. A., 1986, ApJ, 301, 601

WOOSLEY S. E. & WEAVER T. A., 1994, ApJ, 423, 371

VAN DEN HEUVEL E. P. J., BHATTACHARYA D., NOMOTO K. & RAPPAPORT S. A., 1992, A&A, 262, 97

VERMEULEN R. C. et al., 1992, A&A, 270, 204

YUNGELSON L. & LIVIO M., 1998, ApJ, 497, 168

YUNGELSON L. R. & LIVIO M. 1999, in preparation

YUNGELSON L., LIVIO M., TRURAN J. W., TUTUKOV A. & FEDOROVA A. V., 1996, ApJ, 466, 890

YUNGELSON L., LIVIO M., TUTUKOV A. & KENYON S. J., 1995, ApJ, 447, 656

Type Ia supernova progenitors, models, and their cosmological signatures

By P. R U I Z – L A P U E N T E [1,2]
S. B L I N N I K O V [3], R. C A N A L[1], J. M E N D E Z[1,4],
E. S O R O K I N A [5],
A. V I S C O [1], and N. W A L T O N [4]

[1]Department of Astronomy. University of Barcelona, Spain

[2]Max–Planck Institut für Astrophysik, Germany

[3]Institute for Theoretical and Experimental Physics, Moscow, Russia

[4]Isaac Newton Group, La Palma, Spain

[5]Sternberg Astronomical Institute, Moscow, Russia

We overview the candidate binary systems which could lead to Type Ia explosions (SNe Ia) and new ways to discriminate among them. In particular, we show how the change of SNe Ia rates with z indicate the evolutionary timescale of the systems involved. We discuss light curves and spectra from edge–lit detonations and central C+O ignitions. Predictions for both sub–Chandrasekhar and Chandrasekhar explosion models are compared with the observations.

1. Introduction

The interpretation of the spectra of Type Ia supernovae (SNe Ia) at all wavelengths, the location and statistics of these events in their host galaxies, and energetic considerations support the identification of the SNIa phenomenon as the explosion of a carbon–oxygen white dwarf (C–O WD). Moreover, the explosion involves the total disruption of the star, which does not leave any compact object (pulsar, neutron star) as can be inferred from X–ray imaging of the historical Galactic SNe Ia. In order to bring a WD to explosion inducing its thermonuclear explosion, it must be compressed and heated in a process of mass transfer from a second star (Hoyle & Fowler 1960). This places our WD precursors of SNe Ia in binary systems. Although we know that those explosions should take place in binary systems, it has not been possible so far to clearly identify the nature of the system.

In this contribution we address some ways to test the progenitors of SNe Ia. Comparing model spectra and light curves with observations and using the SNe Ia rates at high z to investigate the efficiency of the star–formation process in producing SNe Ia, we hope to learn more about the way those explosions occur and about their progenitors.

2. Favored progenitors

Binary systems that have been investigated as candidates to SNe Ia include: close carbon–oxygen (thereafter, C–O) WD pairs, a C–O WD in a symbiotic system where the companion is an AGB star, a C–O WD plus a He star companion, and a C–O WD with a giant or subgiant (Algol–like systems) or main–sequence star companion (CV–like systems).

2.1. *Mergers*

The binary WD pairs (double degenerate systems) form out of stars from 5 to 8 $M_\odot$, which have gone through the AGB phase and experienced one or two common–envelope phases (Iben & Tutukov 1984; Webbink 1984). The shrinkage of the orbit of the two resulting WDs, as the system emits gravitational wave radiation, ends up in merging of the stars. As the stars approach each other, the tidal forces disrupt the less massive WD. The most massive one is surrounded by a disk (which originated in the disrupted less massive one), it gains mass, and rises its central density and temperature. Close to the Chandrasekhar mass, the accreting WD reaches at its center the conditions for explosive thermonuclear burning. A whole C–O WD plus maybe also some amount of material around are burned. In many cases the amount of C–O material around the WD will be very small and negligible. In some other cases, it might be argued that a thin absorption or emission from the C ions around the WD could be detected in the explosion.

If the smaller WD in the double degenerate system is a He WD, then the outcome of the explosion is different and that is not a good candidate for a SNe Ia explosion. Such a merging would likely result in the formation of a RCrB star (Webbink 1984).

One clear characteristic of the evolutionary path to merging of two WDs is that the mergings peak at about a few times 10^8 yrs after the systems are formed, whereas most of the other candidate systems typically require longer times to explosion (a few 10^9 yr). The time it takes the system to explode is of importance as a possible way to identify the progenitor. Younger systems mean that SNe Ia will be found closer to spiral arms instead of having a spread along the disk characteristic of older stars. Younger systems can also be identified from rates of SNe Ia at high z which follow rather closely the star formation history.

This kind of SNe Ia progenitor should also give rise to a more homogeneous SNe Ia sample than other candidates, as they all explode close to the Chandrasekhar mass. In this scenario the variations seen in the light curves are linked to the outcoming nucleosythesis and expansion of the ejecta: the WD can burn in different modes (deflagration, delayed detonation, pulsating delayed detonation), and incinerate various amounts of material to Fe–peak products. The resulting opacity of the expanding envelope will then be different, and that should account for the variations in light curve decline rate and brightness (Branch et al. 1997).

2.2. *Roche–lobe filling donors*

Among the other types of binary stars which can lead to SNe Ia, close binaries consisting of WDs with subgiant or giant companions (also called Algol–like systems) can produce the growth in mass of the WD until it reaches the explosive condition (Whelan & Iben 1973). The transfer takes place when the subgiant or giant fills its Roche lobe and pours material onto the WD. The material transferred to the WD is H, and depending on the accretion rate the WD will either grow in mass up to the Chandrasekhar limit or ignite a detonation in a outer shell. Cassisi, Iben & Tornambé (1998) find it difficult for those Algol–like systems to reach the Chandrasekhar mass. They found in some cases that a He flash could be produced on top of the WD. If such a He flash is strong enough to compress the WD and induce a central detonation, we have could have in those Algol–type systems a progenitor for sub–Chandrasekhar explosions. The analysis of the difficulty in reaching the Chandrasekhar mass in Algol–type systems differs in the work by Hachisu et al. (1998). The question is not settled at the moment.

Algol–type systems involving the explosion of a very low mass C–O WD were proposed (Ruiz–Lapuente et al. 1993) to be the counterparts for a class of very subluminous SNe Ia (about ten times less luminous than "normal SNe Ia"). The class of these very subluminous SNe Ia was inaugurated by the discovery of SN 1991bg (Filippenko et al. 1992; Leibundgut et al. 1993) and nowadays extends to a handful of events at low z including the recent supernovae SN 1997cn, SN 1998ac (Mazzali et al. 1998; Turatto et al. 1998), whose spectra are better reproduced by the explosion of C–O WD well below the Chandrasekhar mass (about a half of the Chandrasekhar mass). Given that Algol–type systems including a low–mass C–O WD ($M_{WD} \sim$ 0.5–0.6 $M_\odot$) have timescales to explosion of 10^{10} yrs, the suggested progenitors fit well with the fact that those very subluminous SNe Ia are found in elliptical galaxies.

The binaries leading to WD plus subgiant or giant are less massive than those giving rise to mergings: the star which ends up as subgiant or giant companion of the WD can be small (the mass has just to be $M > 0.8 M_\odot$). The mass of the star which forms the WD is in the range 1.8–8 $M_\odot$. This involves evolutionary times of several times 10^9 yr, significantly longer than for the merging of two C+O WDs. The systems which lead to low–mass WDs ($M_{WD} \sim$ 0.5–0.6 $M_\odot$) plus low–mass subgiants would be among the oldest systems responsible for SNe Ia, and certainly present in galaxies which stopped their star formation several billion years ago. With respect to star formation, there will be a delay producing an increase of the local rate of Type Ia supernovae as compared with the rate at high z.

A variant of the same scenario are systems of WD plus a main–sequence companion, i.e, cataclysmic variables. Orbital shrinkage is driven either by magnetic braking or by the emission of gravitational wave radiation. However, the statistics of these binaries, possibly leading to SNe Ia, is too poor to account for the rate of those explosions, neither in our own Galaxy nor in other galaxies (Ruiz–Lapuente, Canal, & Burkert 1997).

2.3. *Other progenitors*

A number of systems have been discarded as likely SNe Ia progenitors: WDs plus He stars (they give quite short timescales to explosions and, therefore, would not produce SNe Ia in elliptical galaxies); WDs in symbiotic systems (accretion from wind does lead, if any, to a small growth in mass of the WD). Tables of conclusions can be found in Ruiz–Lapuente, Canal & Burkert (1997), Branch et al. (1998), and Yungelson & Livio (1998). Recently, Hachisu et al (1996) reevaluated all these progenitors in the light of the effect that a strong accretion–driven wind from the WD has on the evolution of mass growth of the WD. These authors say that, when the formation and modulating effect of the wind is taken into account, a wider range of initial conditions (mass of the stars, separations) produces successful SNe Ia candidates from those binary systems.

3. Signatures of the progenitors

SNe Ia show a degree of variation in brightness and rates of decline of the light curve: from bright and slow decliners to faint and fast decliners, going through normal brightness and normal speed decliners. This spread has been accurately quantified in terms of the δm_{15} parameter (Phillips 1993; Hamuy et al. 1995), stretch factor s (Perlmutter et al. 1998), or Δ factor (Riess et al. 1995). A big step to clarify whether we have various progenitors or just one, involved in the SNe Ia phenomenon, would be to understand this brightness–rate of decline relationship and determine the physical ingredients entering in that correlation. The interest is two–fold: first, only when that correlation will be understood our knowledge of the way SNe Ia explode (sub-Chandrasekhar, Chandrasekhar, one or multiple progenitors) will be complete. Second, that correlation is of key importance for the cosmological use of SNe Ia and we still do not know which are the evolutionary effects that enter into it.

Two alternative explanations are possible for that diversity: WDs explode with different final masses and the spread is due to the diversity in ejecta mass and kinematics, or WDs are basically exploding at the Chandrasekhar mass, and variations in the decline rates are only due to opacity and kinematic effects. In favor of the unique mass stands that most observed SNe Ia seem to cluster around what we know as "normal SNIa" (Branch, Fisher & Nugent 1993), which are SNe Ia in the middle of the variation range and have decline rates (stretch factor) of s around 1. Those are the most frequently observed SNe Ia (not clearly the most intrinsically abundant if there is a subluminous set of SNe Ia, hard to discover at high z). For those supernovae we know at present the intrinsic brightness, since there are Cepheid distance estimates to some of them (SN 1990N, SN 1981B, SN 1989B). Among the "normal SNe Ia class", the intrinsic dispersion is 0.2 mag (Branch, Nugent & Fisher 1997). They seem to be well reproduced at maximum by Chandrasekhar–mass explosions (Nugent et al. 1997), and density diagnostics at late phases suggest that the kinematics/mass of those models is about the observed one (Ruiz–Lapuente 1996). The bolometric light curve can also be reproduced by Chandrasekhar models (Ruiz–Lapuente & Spruit 1998).

The exploration of the alternative hypothesis, sub–Chandrasekar explosions being responsible for the whole SNe Ia class, has also been undertaken. To investigate the whole range of diversity arising in He detonations of sub–Chandrasekhar WDs, Livne & Arnett (1995) (LA95 thereafter) calculated 2-D He detonations of WDs following the accretion of He onto the WD. Depending on the initial mass of the WD, the accreted He layer needed to produce the edge–lit detonation has a different mass. The range of 2-D He detonations in LA95 goes from 0.7 to 1.1 $M_\odot$ WD explosions, with decreasing mass of accreted He as detonator in the surface as the WD mass increases. Similar models were investigated in 1-D by Woosley & Weaver (1994). They are equivalent in the main characteristics of the outcome of the explosions (total mass, ^{56}Ni mass sythesized, kinetic energy).

Being an interesting scenario for SNe Ia, those models have been tested by means of spectral calculations at late phases, bolometric light curves, and light curves in the different colors. From the late phase spectra and late light curves, those 2-D He detonations do not fit "normal SNe Ia" so well as the Chandraskhar explosion models. The ones producing around 0.6 $M_\odot$ of ^{56}Ni have too low densities as compared to what is required (see model 6 by LA95 in Figure 1). The lowest mass model (model 2 by LA95) is too blue and too luminous as compared with the most subluminous SNe Ia (see Figure 2). The tests by Nugent et al. (1997) and Höflich et al. (1997) for the spectra at maximum light of He detonations, also suggest that they are too blue.

Is the bluer signature due to the synthesis of too large an amount of ^{56}Ni plus the small total mass of ejecta in the explosions a killing test? He detonations as presently calculated may well give too blue explosions. However, the explosion model is still unrealistic (Arnett, presented at this meeting), and a 3-D calculation might show some differences in the results.

We have already presented the suggestion of He detonations in low–mass Algol–like systems to be associated with the sublumionous SNe Ia found in elliptical galaxies at low z. In order to produce a He detonation on top of a 0.5 $M_\odot$ WD in the present simulations of He–detonations, accretion of 0.2 $M_\odot$ of He on the WD is required. Thus, the total mass of the WD would be 0.7 $M_\odot$. In the explosion, the external He is mainly converted to ^{56}Ni, and extra quantities of ^{56}Ni are produced near the center from the central ignition ultimately responsible for the WD disruption. The total ^{56}Ni mass is thus high. If there were a way to explode a low–mass WD without accreting as much as

SN 1994D in NGC 4526

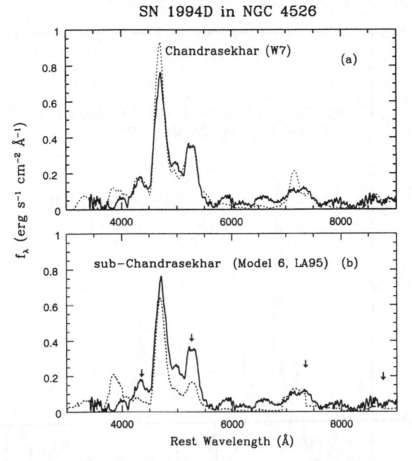

FIGURE 1. Density diagnostics suggest sub–Chandrasekhar He–detonations might not have the right electron density and temperature profiles to reproduce the observations. Comparison of the calculated spectrum of model 6 by Livne & Arnett (1995) and the calculated spectrum of model W7 by Nomoto, Thielemann & Yokoi (1984) with the spectrum of SN 1994D. The solid line shows the observed spectrum and the dashed line the model one.

0.2 $M_\odot$ of He, the result would agree much better with the observations: a simple pure central detonation of such small WDs does indeed account well for the observed features (Ruiz–Lapuente et al. 1993). Alternatively, one could also argue that, in the merging of two WDs, a central structure containing half a Chandrasekhar mass might explode because it gets compressed to high enough densities and temperatures: then we would have another object behaving as those sub–luminous SNe Ia. If that were achieved (such a compression of a low–mass WD to the required central densities is not easy), then a single picture, double degenerate merging, could account for both normal SNe Ia and very subluminous ones.

Following this line of interest on the explosion models, we have tested whether sub–Chandrasekhar explosions are in better agreement with observations than Chandrasekhar–mass explosions. We performed light–curve calculations of several models: as a representative model of central ignition of a Chandrasekhar–mass WD, we take the delayed detonation DD4 (Woosley & Weaver 1994). To further pursue the possibility of low–mass C+O WD explosions, we test the model suggested in 1993 for SN 1991bg (Ruiz–Lapuente

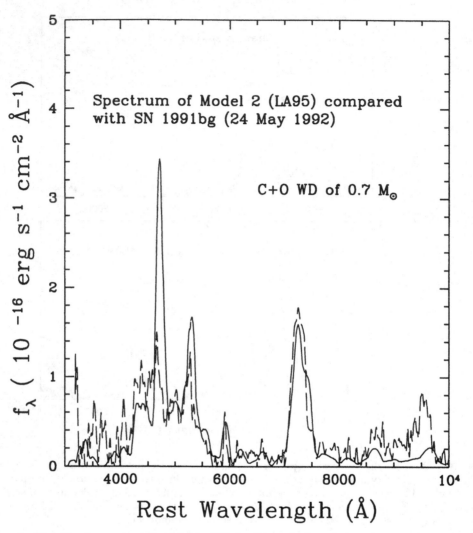

FIGURE 2. The He–detonation of a 0.7 M$_\odot$ WD by Livne & Arnett (1995) compared with the subluminous SNIa SN 1991bg. The solid line shows the model spectrum and the dashed line the observed one.

et al. 1993), and also a low–mass He detonation from Livne & Arnett (1995). The calculations were done with the STELLA code (Blinnikov et al. 1998) and a full account will be given elsewhere (Sorokina et al. 1999, in preparation).

The delayed detonation of a Chadrasekhar–mass WD looks promising for "normal SNe Ia". The results shown here are our first tests of Chandrasekhar/sub–Chandrasekhar explosions through light curves. We expect to perform many more calculations of those models. The light curve results on sub–Chandrasekhar explosions suggest again that He detonation models for the subluminous SNe Ia are too blue and luminous to account for the faint SN 1991bg–like supernovae. A bare small C+O WD with little He accretion would look much more as SN 1991bg–like events.

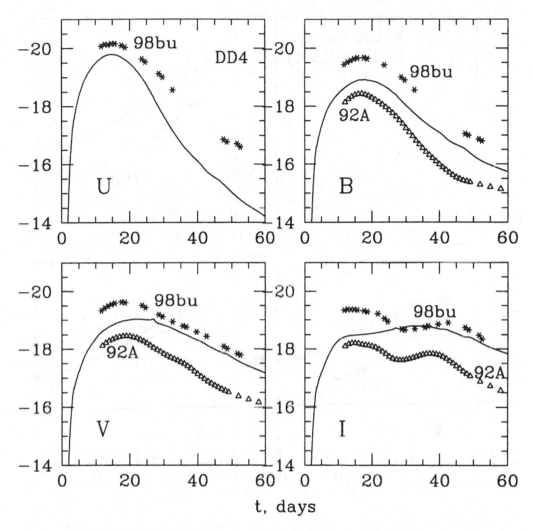

FIGURE 3. Calculated UBVRI light curves of a delayed detonation of a Chadrasekhar mass WD (model DD4). The supernovae SN 1998bu and SN 1992A are shown for a comparison.

4. Efficiencies in SNe Ia production and rates at high z

The study of SNe Ia at high z provides clues on the progenitors (Ruiz–Lapuente, Burkert & Canal 1995; Ruiz–Lapuente, Canal & Burkert 1997; Madau, Della Valle & Panagia 1998). At present, several programmes are providing measurements of the rate of SNe Ia both in the local and in the high–z universes. Cappellaro et al. (1998) present some estimates of the rates from various searches and Hamuy & Pinto (1999) have provided values for the rates from the Calan/Tololo search. At high z, the values so far obtained come from the Supernova Cosmology Project (Pain et al. 1996; 1999 in preparation). The better knowledge of SNe Ia rates at different redshifts, combined with that of the average star formation rate, allows us to investigate whether the efficiency in producing SNe Ia has changed along cosmic time. If metallicity effects do not play any important role, the number of SNe Ia per $M_\odot$ of stars formed should remain nearly constant along the cosmic history. If, on the contrary, metallicity plays an important role

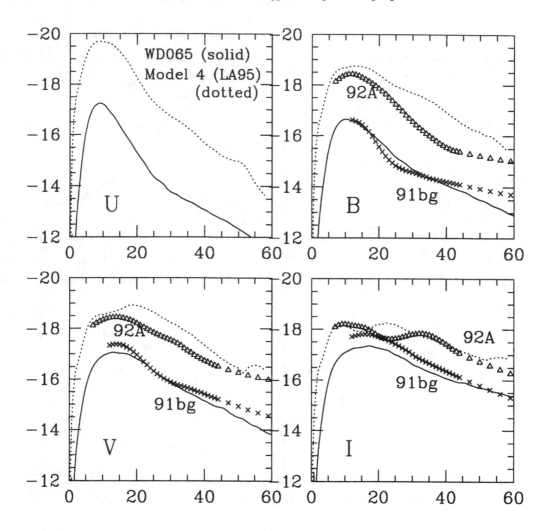

FIGURE 4. Comparison of calculated UBVRI light curves for two models of an explosion of a sub–Chandrasekhar mass WD: model WD065 is a explosion of a 0.65 $M_\odot$ C+O WD (Ruiz–Lapuente et al. 1993) and model 4 (Livne & Arnett 1995) is the explosion of a 0.87 $M_\odot$ WD (mass of the C+O core, 0.7 $M_\odot$). Templates for SN 1992A and for the very subluminous SN 1991bg are also shown.

as, for instance, via the opacity in the wind–modulated accretion model (WD plus Roche-lobe filling subgiant or giant), the number of SNe Ia at low metallicity ($z > 1$) should drastically decrease (Hachisu, Kato & Nomoto 1996). We can define an "instantaneous efficiency" in producing SNe Ia as the rate of SNe Ia (in SN yr^{-1} Mpc^{-3}) divided by star formation rate at a given z (in $M_\odot^{-1}$ Mpc^{-3}). Any change in the efficiency for producing SNe Ia can be a clue as to the progenitor. A drop at $z > 1$ would confirm the metallicity effects suggested by Hachisu et al. (1998). The time that the stars take to reach explosion can also be tested. SNe Ia from merging of WDs having a shorter timescale than those from other systems (about a few times 10^8 yr), would show a similar efficiency at all z, while systems taking a few times 10^9 yr to explode would accumulate towards lower z producing an increase in the "instantaneous SN efficiency" value.

Gallego et al. (1995) estimated the local star formation rate from Hα emission galaxies and derived a value of $\rho_* = 3.7 \times 10^{-2} \ M_\odot \ h^2 \ Mpc^{-3}$. A more recent estimate by Treyer et al. (1998), from a UV–selected galaxy redshift survey, suggests a slightly higher star formation rate. Their estimate of the dust–corrected star formation rate is $\rho_* = 4.3 \times 10^{-2} \ M_\odot \ h^2 \ Mpc^{-3}$.

If we take the SNe Ia rates obtained by Hamuy & Pinto (1998) from the Calan/Tololo survey ($\sim 8.2 \times 10^{-5} \ SNeIa \ yr^{-1} \ h^2 Mpc^{-3}$), and divide it by the local star formation rate, we obtain a local efficiency in SNe Ia production of $\sim 2 \times 10^{-3}$. This means that, locally, every $\sim 500 \ M_\odot$ going into star formation give 1 SNe Ia. Interestingly enough, the efficiency appears to be very similar at high z. When comparing the numbers, we see that the SNe Ia production efficiency (number of SNe Ia per $M_\odot$ going into star formation) is about the same as in the nearby universe. This constancy of the SNe Ia production efficiency if it continues up to $z \sim 1$ would favor the merging of WDs as the main path to SNe Ia explosions.

When one looks to absolute numbers, however, it is easier to account for the rate of SNe Ia at high z with Algol–like systems than with double degenerates, from our current knowledge of the initial distributions (IMF, initial binary mass ratios, initial binary separations) leading to those mergings. SNe Ia counts favor a prolific SNe Ia progenitor. Algol–like systems, given the wider mass ranges of secondaries, can give rise to large SNe Ia numbers, especially if the wind–modulated accretion suggested by Hachisu et al. (1996) widens the allowed ranges in the parameter space of initial masses and separations. Mergings can still do it if we adopt a common envelope parameter and binary distributions of the WDs favoring more mergings than it is usually assumed (Ruiz–Lapuente, Canal & Burkert 1997; Iben et al. 1997; Moran et al. 1997; Yungelson & Livio 1998). With no doubt, this line of research, with upcoming values of the rates at $z \sim 1$, will soon shed light on the progenitor issue.

5. Conclusions

From the complex picture emerging from analysis of SNe Ia by modeling spectra and light curves, as well as by studying the evolution of their rates along cosmic history, we are faced with the following temptative conclusions.

Current He–detonation models give such a balance of kinetic energy/^{56}Ni mass/total ejected mass, that light curves and spectra do not reproduce well normal SNe Ia explosions. On the other hand, the light curves and spectra of models such as delayed detonations of centrally ignited Chandrasekhar–mass WDs look similar to what is observed. Thus, most SNe Ia might come from explosions of WDs close to the Chandrasekhar mass.

Such Chandrasekhar–mass explosions can occur in systems taking about 10^8 yr to explode, i.e. possibly mergers of WDs, or in systems taking several 10^9 yr, of the Algol–type. The timescale can be determined observationally measuring SNe Ia at high z. Given the SNe Ia rate numbers, mergers leading to SNe Ia, if they are the SNe Ia progenitors, should be more frequent than that predicted from current modeling (perhaps a higher common–envelope parameter helps to increase the rates, or maybe the distribution of WD pairs is different from what theory has predicted so far).

To conclude, and as other authors have done, let us add a word on what could happen in respect to the evolution of SNe Ia properties with redshift. As it has been discussed in this conferece, there are some questions which need to be understood about the local and high–z SNe Ia samples. A target of our interest would be to identify at various z the extremes of the SNe Ia distribution (the very subluminuous SNe Ia and the overluminous SNe Ia) and recover the frequency distribution of the members of the SNe Ia family

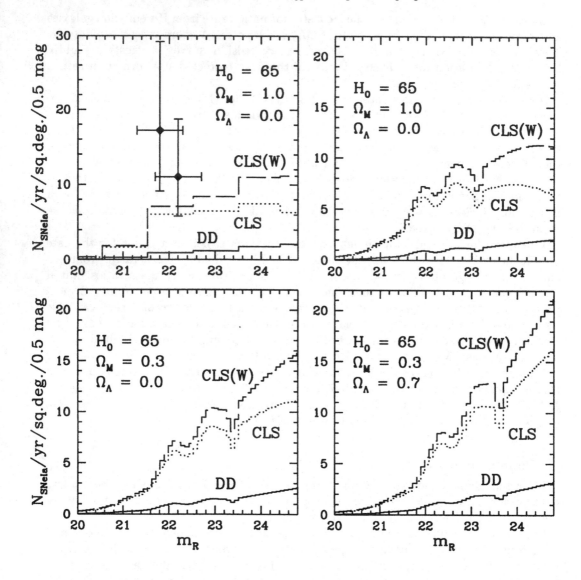

FIGURE 5. Type Ia supernova rates at high z can help us to learn about the SNIa phenomenon. Comparison of observational points (Pain et al. 1996) with predictions (see more in Ruiz–Lapuente & Canal. 1997).

along redshift. Do we have the same spread in the distribution of the local and high–z samples? We detect locally a percentage (~ 15 %) of very subluminous SNe Ia (about ten times less luminous than "normal" SNe Ia). It is necessary to clarify their statistics at high–z. Modeling by various authors has suggested that they might correspond to explosions triggered in low–mass WDs. If they correspond to the suggested low–mass C–O WD explosions in Algol systems, they should disappear back at z larger than ~ 0.4 (since they take almost 10^{10} yr to explode). From the age evolution of SNe Ia progenitors from high z to low z, we could expect that the family at low-z would be wider than at high–z. At high–z we have younger systems and possibly much easier achievers of the Chandrasekhar mass than at low z. That can be making the class of SNe Ia at high z

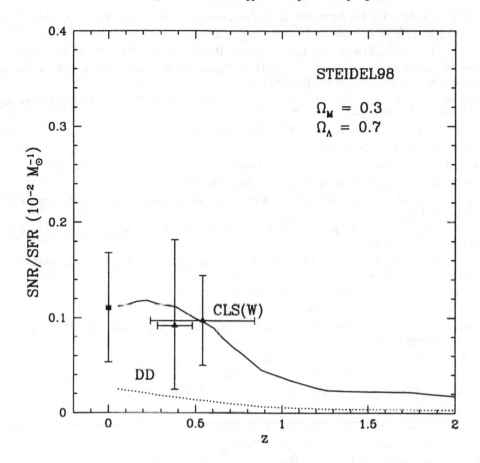

FIGURE 6. Predictions of the ratio SNe Ia to SFR at various z (i.e "instantaneous SNe Ia efficiency"). The merging of two WDs (double degenerate scenario labeled DD) occurs in shorter timescales than the growth in mass of the WD from a Roche–lobe filling companion (Algol–like systems, labeled CLS(W)). Therefore, mergings follow more closely a constant ratio along z–space than Algol–like systems.

cluster more closely around "normal" SNe Ia. Those could be as the "normal" SNe Ia found in our $z \sim 0$ spiral galaxies. Even if the statistical spread of the family at high–z and low–z are different (low-z including more subluminous SNe Ia, for instance), they can still be used accurately for cosmology as long as we do correctly the matching of light curves and magnitudes of high–z SNe Ia to their similars at low z.

Further investigation on our side will include light curve calculations of SNe Ia exploding in different environments and comparisons with a observations at various z.

The work is supported by the INTAS grant "Thermonuclear Supernovae". Part of this work was performed by S.B and E.S at the Stockholm Observatory, which generously allocated computer time.

REFERENCES

BLINNIKOV S. I., EASTMAN R., BARTUNOV O.S, POPOLITOV V.A. & WOOSLEY S.E., 1998, ApJ, 496, 454

BRANCH D., DOGGETT J.B., NOMOTO K. & THIELEMANN F.-K., 1985, ApJ, 348, 647

BRANCH D., FISHER A. & NUGENT P., 1993, AJ, 106, 2383

BRANCH D., LIVIO M., YUNGELSON L.R., BOFFI F.R., BARON E., 1995, PASP, 107, 1019

BRANCH D., NUGENT P. & FISHER A., 1997, in Thermonuclear Supernovae, Eds. P. Ruiz-Lapuente, R. Canal, & J. Isern (Kluwer) 715

CAPPELLARO E., TURATTO M., TSVETKOV D. Y., BARTUNOV O. S., POLLAS C., EVANS R., & HAMUY M., 1997, A&A, 322, 431

CASSISI S., IBEN I.JR, TORNAMBÉ A., 1998, ApJ, 496, 376

FILIPPENKO A. V. et al., 1992, AJ, 104, 1543

GALLEGO J., ZAMORANO J., ARAGON–SALAMANCA A., & REGO M., 1995, ApJ, 361, L1

HACHISU I., KATO M., & NOMOTO K. 1996, ApJ, 470, L97

HAMUY M, PHILLIPS M. M., MAZA J., SUNTZEFF N. B, SCHOMMER R. A., & AVILES R., 1995, AJ, 109, 1

HAMUY M., PHILLIPS M. M., SCHOMMER R. A., SUNTZEFF N. B., MAZA J., & AVILES R., 1996a, AJ, 112, 2399

HAMUY M., PHILLIPS M. M., SUNTZEFF N. B., SCHOMMER R. A., MAZA J., & AVILES R., 1996b, AJ, 112, 2391

HAMUY M & PINTO P. A., 1999, ApJ (submitted)

HACHISU I., KATO M. & NOMOTO K., 1996, ApJ, 470, L97

HÖFLICH P., et al. 1997, in Thermonuclear Supernovae, Eds. P. Ruiz–Lapuente, R. Canal & J. Isern. (Kluwer)

HOYLE F. & FOWLER W.A., 1960, ApJ, 132, 565

IBEN I.,JR., & TUTUKOV A. V., 1984, ApJS, 54, 335

IBEN I. JR., TUTUKOV A. V. & YUNGELSON L. R., 1997, ApJ, 475, 291

LEIBUNDGUT B. et al., 1993, AJ, 105, 301

LIVNE E., & ARNETT D., 1995, ApJ, 452, 62 (LA95)

MADAU P., 1997, (ASP) in press (astro–ph/9707141)

MADAU P. et al., 1996, MNRAS, 283, 1388

MAZZALI P. A., CAPPELLARO E., DANZIGER I. J., TURATTO M., BENETTI S., 1998, ApJ, 499, L49

MORAN C., MARSH T. R. & BRAGAGLIA A., 1997, MNRAS, 288, 538

NUGENT P., PHILLIPS M. M., BARON E., BRANCH D., & HAUSCHILDT P., 1995, ApJ, 455, L147

PAIN R. et al., 1996, ApJ, 473, 356

PERLMUTTER S. et al., 1998, Nature, 391, 51

PHILLIPS M. M. 1993, ApJ, 413, L105

RIESS A. G., PRESS W. H., KIRSHNER R. P., 1995, ApJ ,438, L17

RUIZ–LAPUENTE P., 1996, ApJ, 465, L83

RUIZ–LAPUENTE P., BURKERT A., & CANAL R. 1995, ApJ, 447, L69

RUIZ–LAPUENTE P., CANAL R., & BURKERT A., 1997, in Thermonuclear Supernovae, Eds. P. Ruiz–Lapuente, R. Canal, & J. Isern (Kluwer) 205

RUIZ–LAPUENTE, P. et al., 1993, Nature, 365, 728

RUIZ–LAPUENTE P., KIRSHNER R. P., PHILLIPS M. M., CHALLIS P. M., SCHMIDT B. P., FILIPPENKO A. V., & WHEELER J. C., 1995, ApJ, 439, 60

RUIZ–LAPUENTE P. & SPRUIT H. C., 1998, ApJ, 500, 360

TREYER M. A., ELLIS R. S., MILLARD B., DONAS J. & BRIDGES T. J., 1999, MNRAS (in press)

TURATTO M., PIEMONTE A., BENETTTI S., CAPPELLARO E., MAZZALI P.A., DANZIGER I. J. & PATAT F., 1998, AJ, 116, 2431

TUTUKOV A. V., & YUNGELSON R. L., 1994, MNRAS, 268, 871

WEBBINK R. F., 1984, ApJ, 277, 355

WHELAN J. & IBEN I.JR. 1973, ApJ, 186, 1007

WOOSLEY S. E., & WEAVER T. A. 1994, in Supernovae (Les Houches, Session LIV), Eds. Audouze J. et al. (Elsevier) 63

YUNGELSON L., LIVIO M., TRURAN J. W., TUTUKOV A., & FEDOROVA A., 1996, ApJ, 466, 890

YUNGELSON L. & LIVIO M., 1998, ApJ, 497, 168

Type Ia Supernova Progenitors, Environmental Effects, and Cosmic Supernova Rates

By Ken'ichi NOMOTO[1], Hideyuki UMEDA[1], Izumi HACHISU[2],
Mariko KATO[3], Chiaki KOBAYASHI[1], & Takuji TSUJIMOTO[4]

[1]Department of Astronomy, and Research Center for the Early Universe, University of Tokyo
Tokyo 113-0033
[2]Department of Earth Science and Astronomy, College of Arts and Sciences, University of
Tokyo, Tokyo 153-8902
[3]Department of Astronomy, Keio University, Hiyoshi, Kouhoku-ku, Yokohama 223-8521
[4]National Astronomical Observatory, Mitaka, Tokyo 181-8588

Relatively uniform light curves and spectral evolution of Type Ia supernovae (SNe Ia) have led to the use of SNe Ia as a "standard candle" to determine cosmological parameters, such as the Hubble constant, the density parameter, and the cosmological constant. Whether a statistically significant value of the cosmological constant can be obtained depends on whether the peak luminosities of SNe Ia are sufficiently free from the effects of cosmic and galactic evolutions.

Here we first review the single degenerate scenario for the Chandrasekhar mass white dwarf (WD) models of SNe Ia. We identify the progenitor's evolution and population with two channels: (1) the WD+RG (red-giant) and (2) the WD+MS (near main-sequence He-rich star) channels. In these channels, the strong wind from accreting white dwarfs plays a key role, which yields important age and metallicity effects on the evolution.

We then address the questions whether the nature of SNe Ia depends systematically on environmental properties such as metallicity and age of the progenitor system and whether significant evolutionary effects exist. We suggest that the variation of the carbon mass fraction $X(C)$ in the C+O WD (or the variation of the initial WD mass) causes the diversity of the brightness of SNe Ia. This model can explain the observed dependence of SNe Ia brighness on the galaxy types.

Finally, applying the metallicity effect on the evolution of SN Ia progenitors, we make a prediction of the cosmic supernova rate history as a composite of the supernova rates in different types of galaxies.

1. Introduction

Type Ia supernovae (SNe Ia) are good distance indicators, and provide a promising tool for determining cosmological parameters (e.g., Branch 1998). SNe Ia have been discovered up to $z \sim 1.32$ (Gilliland et al. 1999). Both the Supernova Cosmology Project (Perlmutter et al. 1997, 1999) and the High-z Supernova Search Team (Garnavich et al. 1998; Riess et al. 1998) have suggested a statistically significant value for the cosmological constant.

However, SNe Ia are not perfect standard candles, but show some intrinsic variations in brightness. When determining the absolute peak luminosity of high-redshift SNe Ia, therefore, these analyses have taken advantage of the empirical relation existing between the peak brightness and the light curve shape (LCS). Since this relation has been obtained from nearby SNe Ia only (Phillips 1993; Hamuy et al. 1995; Riess et al. 1995), it is important to examine whether it depends systematically on environmental properties such as metallicity and age of the progenitor system.

High-redshift supernovae present us very useful information, not only to determine cosmological parameters but also to put constraints on the star formation history in the universe. They have given the SN Ia rate at $z \sim 0.5$ (Pain 1999) but will provide the SN Ia rate history over $0 < z < 1$. With the Next Generation Space Telescope, both SNe Ia and SNe II will be observed through $z \sim 4$. It is useful to provide a prediction of cosmic supernova rates to constrain the age and metallicity effects of the SN Ia progenitors.

SNe Ia have been widely believed to be a thermonuclear explosion of a mass-accreting white dwarf (WD) (e.g., Nomoto et al. 1997a for a review). However, the immediate progenitor binary systems have not been clearly identified yet (Branch et al. 1995). In order to address the above questions regarding the nature of high-redshift SNe Ia, we need to identify the progenitors systems and examine the "evolutionary" effects (or environmental effects) on those systems.

In §2, we summarize the progenitors' evolution where the strong wind from accreting WDs plays a key role (Hachisu, Kato, & Nomoto 1996, 1999, hereafter HKN96, HKN99; Hachisu, Kato, Nomoto & Umeda 1999, HKNU99). In §3, we addresses the issue of whether a difference in the environmental properties is at the basis of the observed range of peak brightness (Umeda et al. 1999b). In §4, we make a prediction of the cosmic supernova rate history as a composite of the different types of galaxies (Kobayashi et al. 1999).

2. Progenitor's evolution and the white dwarf wind

There exist two models proposed as progenitors of SNe Ia: 1) the Chandrasekhar mass model, in which a mass-accreting carbon-oxygen (C+O) WD grows in mass up to the critical mass $M_{Ia} \simeq 1.37 - 1.38 M_\odot$ near the Chandrasekhar mass and explodes as an SN Ia (e.g., Nomoto et al. 1984, 1994), and 2) the sub-Chandrasekhar mass model, in which an accreted layer of helium atop a C+O WD ignites off-center for a WD mass well below the Chandrasekhar mass (e.g., Arnett 1996). The early time spectra of the majority of SNe Ia are in excellent agreement with the synthetic spectra of the Chandrasekhar mass models, while the spectra of the sub-Chandrasekhar mass models are too blue to be comparable with the observations (Höflich & Khokhlov 1996; Nugent et al. 1997).

For the evolution of accreting WDs toward the Chandrasekhar mass, two scenarios have been proposed: 1) a double degenerate (DD) scenario, i.e., merging of double C+O WDs with a combined mass surpassing the Chandrasekhar mass limit (Iben & Tutukov 1984; Webbink 1984), and 2) a single degenerate (SD) scenario, i.e., accretion of hydrogen-rich matter via mass transfer from a binary companion (e.g., Nomoto 1982; Nomoto et al. 1994). The issue of DD vs. SD is still debated (e.g., Branch et al. 1995), although theoretical modeling has indicated that the merging of WDs leads to the accretion-induced collapse rather than SN Ia explosion (Saio & Nomoto 1985, 1998; Segretain et al. 1997).

In the SD Chandrasekhar mass model for SNe Ia, a WD explodes as a SN Ia only when its rate of the mass accretion ($\dot{M}$) is in a certain narrow range (e.g., Nomoto 1982; Nomoto & Kondo 1991). In particular, if $\dot{M}$ exceeds the critical rate $\dot{M}_b$ in Eq.(2.1) below, the accreted matter extends to form a common envelope (Nomoto et al. 1979). This difficulty has been overcome by the WD wind model (see below). For the actual binary systems which grow the WD mass (M_{WD}) to M_{Ia}, the following two systems are appropriate. One is a system consisting of a mass-accreting WD and a lobe-filling, more massive, slightly evolved main-sequence (MS) or sub-giant star (hereafter "WD+MS system"). The other system consists of a WD and a lobe-filling, less massive, red-giant (hereafter "WD+RG system").

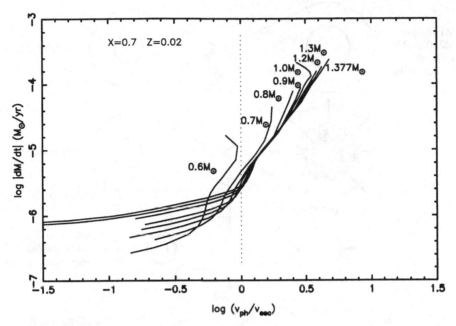

FIGURE 1. Ratio of the photospheric velocity to the escape velocity there v_{ph}/v_{esc} is plotted against the decreasing rate of the envelope mass for WDs with masses of $0.6M_\odot$, $0.7M_\odot$, $0.8M_\odot$, $0.9M_\odot$, $1.0M_\odot$, $1.2M_\odot$, $1.3M_\odot$, and $1.377M_\odot$. We regard the wind as "strong" when the photospheric velocity exceeds the escape velocity there, i.e., $v_{ph} > v_{esc}$. If not, it is regarded as "weak."

2.1. *White dwarf winds*

Optically thick WD winds are driven when the accretion rate $\dot{M}$ exceeds the critical rate $\dot{M}_b$. Here $\dot{M}_b$ is the rate at which steady burning can process the accreted hydrogen into helium as

$$\dot{M}_b \approx 0.75 \times 10^{-6} \left(\frac{M_{WD}}{M_\odot} - 0.40 \right) M_\odot \; yr^{-1}. \qquad (2.1)$$

With such a rapid accretion, the WD envelope expands to $R_{ph} \sim 0.1 R_\odot$ and the photospheric temperature decreases below $\log T_{ph} \sim 5.5$. Around this temperature, the shoulder of the strong peak of OPAL Fe opacity (Iglesias & Rogers 1993) drives the radiation-driven wind (HKN96; HKN99). We plot the ratio v_{ph}/v_{esc} between the photospheric velocity and the escape velocity at the photosphere in Figure 1 against the mass transfer rate. We call the wind *strong* when $v_{ph} > v_{esc}$. When the wind is strong, $v_{ph} \sim 1000$ km s^{-1} being much faster than the orbital velocity.

If the wind is sufficiently strong, the WD can avoid the formation of a common envelope and steady hydrogen burning increases its mass continuously at a rate $\dot{M}_b$ by blowing the extra mass away in a wind. When the mass transfer rate decreases below this critical value, optically thick winds stop. If the mass transfer rate further decreases below ~ 0.5 $\dot{M}_b$, hydrogen shell burning becomes unstable to trigger very weak shell flashes but still burns a large fraction of accreted hydrogen.

The steady hydrogen shell burning converts hydrogen into helium atop the C+O core and increases the mass of the helium layer gradually. When its mass reaches a certain value, weak helium shell flashes occur. Then a part of the envelope mass is blown off but a large fraction of He can be burned to C+O (Kato & Hachisu 1999) to increase the WD

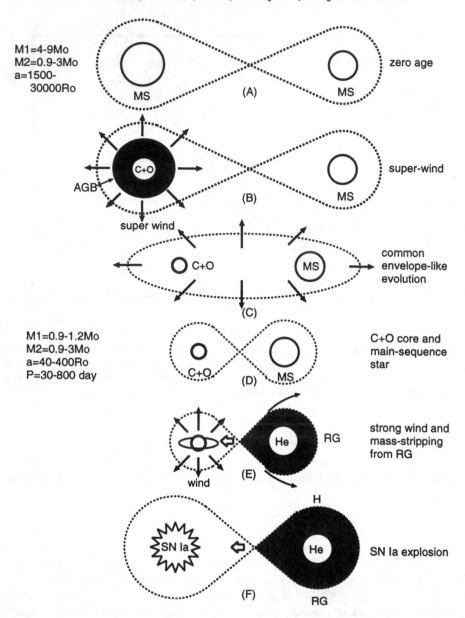

FIGURE 2. An illustration of the WD+RG (symbiotic) channel to Type Ia supernovae.

mass. In this way, Thus strong winds from the accreting WD play a key role to increase the WD mass to $M_{\rm Ia}$.

2.2. *WD+RG system*

This is a symbiotic binary system consisting of a white dwarf (WD) and a low mass red-giant (RG). A full evolutionary path of the WD+RG system from the zero age main-sequence stage (*stage A*) to the SN Ia explosion (*stage F*) is as follows.

(A) Zero age main-sequence.

(B) The primary has evolved first to become an asymptotic giant branch (AGB) star and blows a slow wind (or a super wind) at the end of the AGB evolution.

(C) If the superwind from the AGB star is as fast as or slower than the orbital velocity, the wind outflowing from the system takes away the orbital angular momentum effectively. As a result the wide binary shrinks greatly (by about a factor of ten or more) to become a close binary. This is a similar process to the common envelope evolution.

(D) Then the AGB star undergoes a common envelope evolution. The AGB star forms a C+O WD, and the initial secondary remains a main-sequence star (MS).

(E) The initial secondary evolves to a red-giant (RG) forming a helium core and fills up its inner critical Roche lobe. Mass transfer begins. The WD component blows a strong wind and the winds can stabilize the mass transfer even if the RG component has a deep convective envelope.

(F) The WD component has grown in mass to M_{Ia} and explodes as a Type Ia supernova.

The occurrence frequency of SNe Ia through this channel is much larger than the earlier scenario, because of the following two evolutionary processes, which have not considered before.

(1) Because of the AGB wind at stage C, the WD + RG close binary can form from a wide binary even with such a large initial separation as $a_i \lesssim 40,000 R_\odot$. Our earlier estimate (HKN96) is constrained by $a_i \lesssim 1,500 R_\odot$.

(2) When the RG fills its inner critical Roche lobe, the WD undergoes rapid mass accretion and blows a strong optically thick wind. Our earlier analysis has shown that the mass transfer is stabilized by this wind only when the mass ratio of RG/WD is smaller than 1.15. Our new finding is that the WD wind can strip mass from the RG envelope, which could be efficient enough to stabilize the mass transfer even if the RG/WD mass ratio exceeds 1.15. If this mass-stripping effect is strong enough, though its efficiency η_{eff} is subject to uncertainties, the symbiotic channel can produce SNe Ia for a much (ten times or more) wider range of the binary parameters than our earlier estimation.

With the above two new effects (1) and (2), the WD+RG (symbiotic) channel can account for the inferred rate of SNe Ia in our Galaxy. The immediate progenitor binaries in this symbiotic channel to SNe Ia may be observed as symbiotic stars, luminous supersoft X-ray sources, or recurrent novae like T CrB or RS Oph, depending on the wind status.

2.3. *WD+MS system*

In this scenario, a C+O WD is originated, not from an AGB star with a C+O core, but from a red-giant star with a helium core of $\sim 0.8 - 2.0 M_\odot$. The helium star, which is formed after the first common envelope evolution, evolves to form a C+O WD of $\sim 0.8 - 1.1 M_\odot$ with transferring a part of the helium envelope onto the secondary main-sequence star.

As an example for this system, let us consider a pair of $7 M_\odot + 2.5 M_\odot$ with the initial separation of $a_i \sim 50 - 600 R_\odot$. The binary evolves to SN Ia through the following stages (stages A-F in Figure 3 and G in HKNU99):

(1) stage A-C: When the mass of the helium core grows to $1.0 M_\odot < M_{1,\mathrm{He}} < 1.4 M_\odot$, the primary fills its Roche lobe and the binary undergoes a common envelope evolution.

(2) stage C-D: After the common envelope evolution, the system consists of a helium star and a main-sequence star with a relatively compact separation of $a_f \sim 3 - 40 R_\odot$ and $P_{\mathrm{orb}} \sim 0.4 - 20$ d.

(3) stage D: The helium star contracts and ignites central helium burning to become a helium main-sequence star. The primary stays at the helium main-sequence for $\sim 1 \times 10^7$ yr.

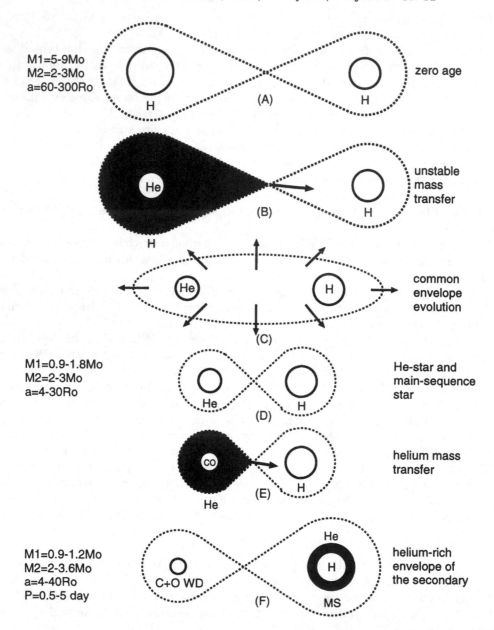

M1=5-9Mo
M2=2-3Mo
a=60-300Ro

(A)   zero age

He   unstable
                                  mass
(B)   transfer

He   common
                                  envelope
(C)   evolution

M1=0.9-1.8Mo
M2=2-3Mo
a=4-30Ro

He   H   He-star and
                                  main-sequence
(D)   star

co   helium mass
                                  transfer
He
(E)

M1=0.9-1.2Mo
M2=2-3.6Mo
a=4-40Ro
P=0.5-5 day

C+O WD   helium-rich
He   H   envelope of
(F)   MS   the secondary

FIGURE 3. An illustration of the WD+MS channel to Type Ia supernovae through the common envelope evolution to the helium matter transfer.

(4) stage E: After helium exhaustion, a carbon-oxygen core develops. When the core mass reaches $0.9 - 1.0 M_\odot$, the helium star evolves to a red-giant and fills again its inner critical Roche lobe. Almost pure helium is transferred to the secondary because the mass transfer is stable for the mass ratio $q = M_1/M_2 < 0.79$. The mass transfer rate is as large as $\sim 1 \times 10^{-5} M_\odot$ yr^{-1} but the mass-receiving main-sequence star ($\sim 2 - 3 M_\odot$) does not expand for such a low rate.

(5) stage F: The secondary has received $0.1 - 0.4 M_\odot$ (almost) pure helium and, as a result, it becomes a helium-rich star as observed in the recurrent nova U Sco. The

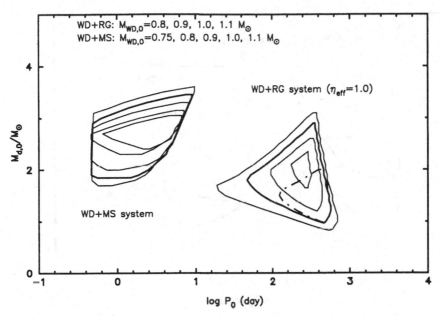

FIGURE 4. The region to produce SNe Ia in the $\log P_0 - M_{d,0}$ plane for five initial white dwarf masses of $0.75 M_\odot$, $0.8 M_\odot$, $0.9 M_\odot$, $1.0 M_\odot$ (heavy solid line), and $1.1 M_\odot$. The region of $M_{WD,0} = 0.7 M_\odot$ almost vanishes for both the WD+MS and WD+RG systems, and the region of $M_{WD,0} = 0.75 M_\odot$ vanishes for the WD+RG system. Here, we assume the stripping efficiency of $\eta_{eff} = 1$. For comparison, we show only the region of $M_{WD,0} = 1.0 M_\odot$ for a much lower efficiency of $\eta_{eff} = 0.3$ by a dash-dotted line.

primary becomes a C+O WD. The separation and thus the orbital period gradually increases during the mass transfer phase. The final orbital period becomes $P_{orb} \sim 0.5-40$ d.

(6) stage G: The white dwarf accretes hydrogen-rich, helium-enhanced matter from a lobe-filling, slightly evolved companion at a critical rate and blows excess matter in the wind. The white dwarf grows in mass to M_{Ia} and explodes as an SN Ia.

This evolutionary path provides a much wider channel to SNe Ia than previous scenarios. A part of the progenitor systems are identified as the luminous supersoft X-ray sources (van den Heuvel et al. 1992) during steady H-burning (but without wind to avoid extinction), or the recurrent novae like U Sco if H-burning is weakly unstable. Actually these objects are characterized by the accretion of helium-rich matter.

2.4. *Realization Frequency*

For an immediate progenitor system WD+RG of SNe Ia, we consider a close binary initially consisting of a C+O WD with $M_{WD,0} = 0.6 - 1.2 M_\odot$ and a low-mass red-giant star with $M_{RG,0} = 0.7 - 3.0 M_\odot$ having a helium core of $M_{He,0} = 0.2 - 0.46 M_\odot$ (stage E). The initial state of these immediate progenitors is specified by three parameters, i.e., $M_{WD,0}$, $M_{RG,0} = M_{d,0}$, and the initial orbital period P_0 ($M_{He,0}$ is determined if P_0 is given).

We follow binary evolutions of these systems and obtain the parameter range(s) which can produce an SN Ia. In Figure 4, the region enclosed by the thin solid line produces SNe Ia for several cases of the initial WD mass, $M_{WD,0} = 0.75 - 1.1\ M_\odot$. For smaller $M_{WD,0}$,

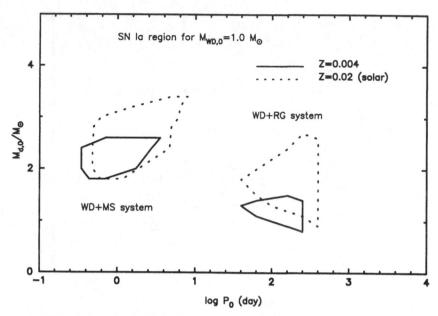

FIGURE 5. The regions of SNe Ia is plotted in the initial orbital period vs. the initial companion mass diagram for the initial WD mass of $M_{WD,0} = 1.0 M_\odot$. The dashed and solid lines represent the cases of solar abundance ($Z = 0.02$) and much lower metallicity of $Z = 0.004$, respectively. The left and the right regions correspond to the WD+MS and the WD+RG systems, respectively.

the wind is weaker, so that the SN Ia region is smaller. The regions of $M_{WD,0} = 0.6 M_\odot$ and $0.7 M_\odot$ vanish for both the WD+MS and WD+RG systems.

In the outside of this region, the outcome of the evolution at the end of the calculations is not an SN Ia but one of the followings:

(i) Formation of a common envelope for too large M_d or $P_0 \sim$ day, where the mass transfer is unstable at the beginning of mass transfer.

(ii) Novae or strong hydrogen shell flash for too small $M_{d,0}$, where the mass transfer rate becomes below 10^{-7} $M_\odot$ yr^{-1}.

(iii) Helium core flash of the red giant component for too long P_0, where a central helium core flash ignites, i.e., the helium core mass of the red-giant reaches $0.46 M_\odot$.

(iv) Accretion-induced collapse for $M_{WD,0} > 1.2 M_\odot$, where the central density of the WD reaches $\sim 10^{10}$ g cm^{-3} before heating wave from the hydrogen burning layer reaches the center. As a result, the WD undergoes collapse due to electron capture without exploding as an SN Ia (Nomoto & Kondo 1991).

It is clear that the new region of the WD+RG system is not limited by the condition of $q < 1.15$, thus being ten times or more wider than the region of HKN96's model (depending on the the stripping efficiency of η_{eff}).

The WD+MS progenitor system can also be specified by three initial parameters: the initial C+O WD mass $M_{WD,0}$, the mass donor's initial mass $M_{d,0}$, and the orbital period P_0. For $M_{WD,0} = 1.0 M_\odot$, the region producing an SN Ia is bounded by $M_{d,0} = 1.8 - 3.2 M_\odot$ and $P_0 = 0.5 - 5$ d as shown by the solid line in Figure 4. The upper and lower bounds are respectively determined by the common envelope formation (i) and nova-like explosions (ii) as above. The left and right bounds are determined by the minimum and maximum radii during the main sequence of the donor star (HKNU99).

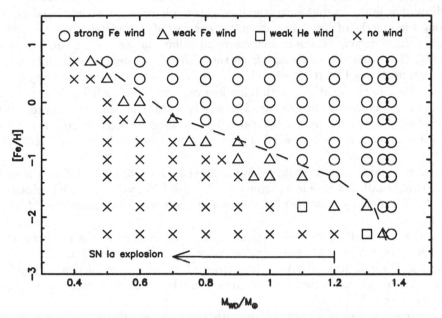

FIGURE 6. WD mass vs. metallicity diagram showing the metallicity dependence of optically thick winds We regard the wind as "strong" if the wind velocity at the photosphere exceeds the escape velocity but "weak" if the wind velocity is lower than the escape velocity. The term of "He" or "Fe" wind denotes that the wind is accelerated by the peak of iron lines near $\log T(K) \sim 5.2$ or of helium lines near $\log T(K) \sim 4.6$. The dashed line indicates the demarcation between the "strong" wind and the "weak" wind.

We estimate the rate of SNe Ia originating from these channels in our Galaxy by using equation (1) of Iben & Tutukov (1984). The realization frequencies of SNe Ia through the WD+RG and WD+MS channels are estimated as ~ 0.0017 yr^{-1} (WD+RG) and ~ 0.001 yr^{-1} (WD+MS), respectively. The total SN Ia rate of the WD+MS/WD+RG systems becomes ~ 0.003 yr^{-1}, which is close enough to the inferred rate of our Galaxy.

2.5. *Low metallicity inhibition of type Ia supernovae*

In the above SN Ia progenitor model, the accreting WD blows a strong wind to reach the Chandrasekhar mass limit. If the iron abundance of the progenitors is as low as [Fe/H]$\lesssim -1$, then the wind is too weak for SNe Ia to occur.

In this model, an interesting metallicity effect has been found (Kobayashi et al. 1998; Hachisu & Kato 1999). The wind velocity is higher for larger M_{WD} and larger Fe/H because of higher luminosity and larger opacity, respectively. In order to blow sufficiently strong wind ($v_w > v_{esc}$), M_{WD} should exceed a certain mass M_w (Fig. 1). As seen from the dashed line in Figure 6, M_w is larger for lower metallicity; e.g., $M_w = 0.65$, 0.85, and 0.95 $M_\odot$ for $Z = 0.02$, 0.01, and 0.004, respectively. In order for a WD to grow its mass at $\dot{M} > \dot{M}_b$, its initial mass $M_{WD,0}$ should exceed M_w. In other words, M_w is the metallicity-dependent minimum $M_{WD,0}$ required for a WD to become an SN Ia.

2.6. *Possible detection of hydrogen*

In our scenario, the WD winds form a circumstellar envelope around the binary systems prior to the explosion, which may emit X-rays, radio, and Hα lines by shock heating when the ejecta collide with the circumstellar envelope. The mass accretion rate in the present models is still as high as 1×10^{-6} $M_\odot$ yr^{-1} for some of the white dwarfs near the Chandrasekhar limit, so that such a white dwarf explode in the strong wind phase.

Our strong wind model of case P1 predicts the presence of circumstellar matter around the exploding white dwarf. Whether such a circumstellar matter is observable depends on its density. The wind mass loss rate from the white dwarf near the Chandrasekhar limit is as high as $\dot{M} \sim 1 \times 10^{-8}$ -1×10^{-7} $M_\odot$ yr^{-1} and the wind velocity is $v = 1000$ km s^{-1}. Despite the relatively high mass loss rate, the circumstellar density is not so high because of the high wind velocity. For steady wind, the density is expressed by $\dot{M}/v$ ($= 4\pi r^2 \rho$). Normalized by the typical red-giant wind velocity of 10 km s^{-1}, the density measure of our white dwarf wind is given as $\dot{M}/v_{10} \sim 1 \times 10^{-10}$ -1×10^{-9} $M_\odot$ yr^{-1}, where $v_{10} = v/10$ km s^{-1}.

Behind the red-giant, matter stripped from the red-giant component forms a much dense circumstellar tail. Its rate is as large as $\sim 1 \times 10^{-7} M_\odot$ yr^{-1} with the velocity of ~ 100 km s^{-1}. The density measure of the dense red-giant wind thus formed is $\dot{M}/v_{10} \sim 1 \times 10^{-8}$ $M_\odot$ yr^{-1}.

Further out, circumstellar matter is produced from the wind from the red-giant companion, which is too far away to cause significant circumstellar interaction.

For some cases, winds from the WD have stopped before the explosion. Therefore, circumstellar matter is dominated by the wind from the red-giant companion whose velocity is as low as ~ 10 km s^{-1}.

At SN Ia explosion, ejecta would collide with the circumstellar matter, which produces shock waves propagating both outward and inward. At the shock front, particle accelerations take place to cause radio emissions. Hot plasmas in the shocked materials emit thermal X-rays. The circumstellar matter ahead of the shock is ionized by X-rays and produce recombination Hα emissions (Cumming et al. 1996). Such an interaction has been observed in Type Ib, Ic, and II supernovae, most typically in SN 1993J (e.g., Suzuki & Nomoto 1995 and references therein).

For SNe Ia, several attempts have been made to detect the above signature of circumstellar matter. There has been no radio and X-ray detections so far. The upper limit set by X-ray observations of SN 1992A is $\dot{M}/v_{10} = (2-3) \times 10^{-6}$ $M_\odot$ yr^{-1} (Schlegel & Petre 1993). Radio observations of SN 1986G have provided the most strict upper limit to the circumstellar density as $\dot{M}/v_{10} = 1 \times 10^{-7}$ $M_\odot$ yr^{-1} (Eck et al. 1995). This is still $10-100$ times higher than the density predicted for the white dwarf wind. If the WD wind has ceased and the wind mass loss rate from the red-giant is significantly higher than 1×10^{-7} $M_\odot$ yr^{-1}, radio detection could be possible for very nearby SNe Ia as close as SN 1986G. (Note also that SN 1986G is not a typical SN Ia but a subluminous SN Ia.)

For Hα emissions, Branch et al. (1983) noted a small, narrow emission feature at the rest wavelength of Hα, which is blueshifted by 1800 km s^{-1} from the local interstellar Ca II absorption. Though this feature was not observed 5 days later, such high velocity hydrogen is expected from the white dwarf wind model. For SN 1994D, Cumming et al. (1996) obtained the upper limit of $\dot{M}/v_{10} = 6 \times 10^{-6}$ $M_\odot$ yr^{-1}. Further attempts to detect Hα emissions are highly encouraged.

3. The origin of diversity of SNe Ia and environmental effects

There are some observational indications that SNe Ia are affected by their environment. The most luminous SNe Ia seem to occur only in spiral galaxies, while both spiral and elliptical galaxies are hosts for dimmer SNe Ia. Thus the mean peak brightness is dimmer in ellipticals than in spiral galaxies (Hamuy et al. 1996). The SNe Ia rate per

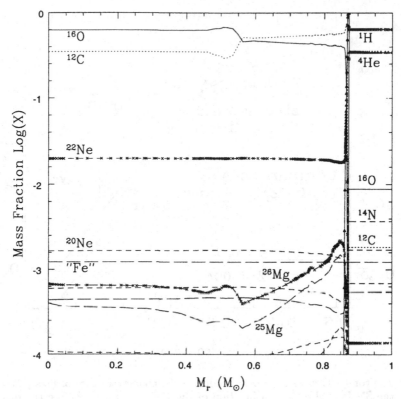

FIGURE 7. Abundances in mass fraction in the inner core of the 6 $M_\odot$ star for $Y = 0.2775$ and $Z = 0.02$ at the end of the second dredge-up.

unit luminosity at the present epoch is almost twice as high in spirals as in ellipticals (Cappellaro et al. 1997). Moreover, Wang et al. (1997) and Riess et al. (1999) found that the variation of the peak brightness for SNe located in the outer regions in galaxies is smaller.

Höflich et al. (1998, 1999) examined how the initial composition of the WD (metallicity and the C/O ratio) affects the observed properties of SNe Ia. Umeda et al. (1999a) obtained the C/O ratio as a function of the main-sequence mass and metallicity of the WD progenitors. Umeda et al. (1999b) suggested that the variation of the C/O ratio is the main cause of the variation of SNe Ia brightness, with larger C/O ratio yielding brighter SNe Ia. We will show that the C/O ratio depends indeed on environmental properties, such as the metallicity and age of the companion of the WD, and that our model can explain most of the observational trends discussed above. We then make some predictions about the brightness of SN Ia at higher redshift.

3.1. *C/O ratio in WD progenitors*

In this section we discuss how the C/O ratio in the WD depends on the metallicity and age of the binary system. The C/O ratio in C+O WDs depends primarily on the main-sequence mass of the WD progenitor and on metallicity.

We calculated the evolution of intermediate-mass $(3 - 9M_\odot)$ stars for metallicity Z=0.001 – 0.03. In the ranges of stellar masses and Z considered in this paper, the most important metallicity effect is that the radiative opacity is smaller for lower Z. Therefore, a star with lower Z is brighter, thus having a shorter lifetime than a star with

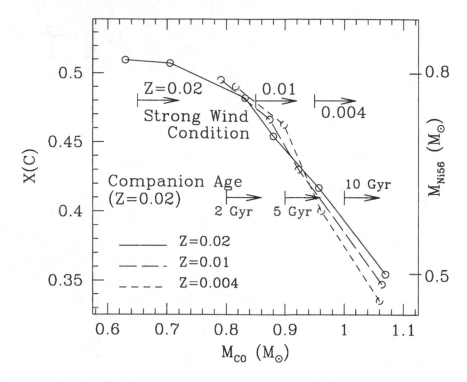

FIGURE 8. The total ^{12}C mass fraction included in the convective core of mass, $M = 1.14 M_\odot$, just before the SN Ia explosion as a function of the C+O core mass before the onset of mass accretion, M_{CO}. The lower bounds of M_{CO} obtained from the age effects and the conditions for strong wind to blow are also shown by arrows.

the same mass but higher Z. In this sense, the effect of reducing metallicity for these stars is almost equivalent to increasing a stellar mass.

For stars with larger masses and/or smaller Z, the luminosity is higher at the same evolutionary phase. With a higher nuclear energy generation rate, these stars have larger convective cores during H and He burning, thus forming larger He and C-O cores.

As seen in Figure 7, the central part of these stars is oxygen-rich. The C/O ratio is nearly constant in the innermost region, which was a convective core during He burning. Outside this homogeneous region, where the C-O layer grows due to He shell burning, the C/O ratio increases up to C/O $\gtrsim$ 1; thus the oxygen-rich core is surrounded by a shell with C/O $\gtrsim$ 1. In fact this is a generic feature in all models we calculated. The C/O ratio in the shell is C/O $\simeq$ 1 for the star as massive as $\sim 7 M_\odot$, and C/O $>$ 1 for less massive stars.

When a progenitor reaches the critical mass for the SNe Ia explosion, the central core is convective up to around 1.1 $M_\odot$. Hence the relevant C/O ratio is between the central value before convective mixing and the total C/O of the whole WD. Using the results from the C6 model (Nomoto et al. 1984), we assume that the convective region is 1.14 $M_\odot$ and for simplicity, C/O = 1 outside the C-O core at the end of second dredge-up. Then we obtain the C/O ratio of the inner part of the SNe Ia progenitors (Fig. 8).

From this figure we find three interesting trends. First, while the central C/O is a complicated function of stellar mass (Umeda et al. 1999a), as shown here the C/O ratio in the core before SNe Ia explosion is a decreasing monotonic function of mass. The central C/O ratio at the end of second dredge-up decreases with mass for $M_{ms} \gtrsim 5M_\odot$, while the ratio increases with mass for $M_{ms} \gtrsim 4M_\odot$; however, the convective core mass during He burning is smaller for a less massive star, and the C/O ratio during shell He burning is larger for smaller C+O core. Hence, when the C/O ratio is averaged over 1.1 $M_\odot$ the C/O ratio decreases with mass. Second, as shown in Umeda et al. 1999, although the C/O ratio is a complicated function of metallicity and mass, the metallicity dependence is remarkably converged when the ratio is seen as a function of the C+O core mass (M_{CO}) instead of the initial main sequence mass.

According to the evolutionary calculations for $3-9$ $M_\odot$ stars by Umeda et al. (1999a), the C/O ratio and its distribution are determined in the following evolutionary stages of the close binary.

(1) At the end of central He burning in the $3-9$ $M_\odot$ primary star, C/O< 1 in the convective core. The mass of the core is larger for more massive stars.

(2) After central He exhaustion, the outer C+O layer grows via He shell burning, where C/O$\gtrsim$1 (Umeda et al. 1999a).

(3a) If the primary star becomes a red giant (case C evolution; e.g. van den Heuvel 1994), it then undergoes the second dredge-up, forming a thin He layer, and enters the AGB phase. The C+O core mass, M_{CO}, at this phase is larger for more massive stars. For a larger M_{CO} the total carbon mass fraction is smaller.

(3b) When it enters the AGB phase, the star greatly expands and is assumed here to undergo Roche lobe overflow (or a super-wind phase) and to form a C+O WD. Thus the initial mass of the WD, $M_{WD,0}$, in the close binary at the beginning of mass accretion is approximately equal to M_{CO}.

(4a) If the primary star becomes a He star (case BB evolution), the second dredge-up in (3a) corresponds to the expansion of the He envelope.

(4b) The ensuing Roche lobe overflow again leads to a white dwarf of mass $M_{WD,0} = M_{CO}$.

(5) After the onset of mass accretion, the WD mass grows through steady H burning and weak He shell flashes, as described in the WD wind model. The composition of the growing C+O layer is assumed to be C/O=1.

(6) The WD grows in mass and ignites carbon when its mass reaches $M_{Ia} = 1.367M_\odot$, as in the model C6 of Nomoto et al. (1984). Because of strong electron-degeneracy, carbon burning is unstable and grows into a deflagration for a central temperature of 8×10^8 K and a central density of 1.47×10^9 g cm^{-3}. At this stage, the convective core extends to $M_r = 1.14M_\odot$ and the material is mixed almost uniformly, as in the C6 model.

In Figure 8, we show the carbon mass fraction $X(C)$ in the convective core of this pre-explosive WD, as a function of metallicity (Z) and initial mass of the WD before the onset of mass accretion, M_{CO}. Figure 8 reveals that: 1) $X(C)$ is smaller for larger $M_{CO} \simeq M_{WD,0}$. 2) The dependence of $X(C)$ on metallicity is small when plotted against M_{CO}, even though the relation between M_{CO} and the initial stellar mass depends sensitively on Z (Umeda et al. 1999a).

3.2. *Brightness of SNe Ia and the C/O ratio*

In the Chandrasekhar mass models for SNe Ia, the brightness of SNe Ia is determined mainly by the mass of ^{56}Ni synthesized (M_{Ni56}). Observational data suggest that M_{Ni56} for most SNe Ia lies in the range $M_{Ni56} \sim 0.4 - 0.8M_\odot$ (e.g. Mazzali et al. 1998). This range of M_{Ni56} can result from differences in the C/O ratio in the progenitor WD as follows.

In the deflagration model, a larger C/O ratio leads to the production of more nuclear

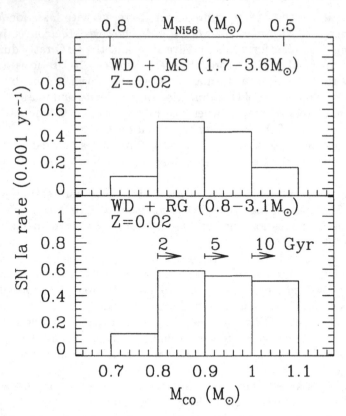

FIGURE 9. SN Ia frequency for a galaxy of mass $2 \times 10^{11} M_\odot$ as a function of M_{CO} for Z=0.02. For the WD+RG system, constraints from the companion's age are shown by the arrows. SNe Ia from the WD+MS system occur in spirals but not in ellipticals because of the age effect. M_{CO} and M_{Ni56} is assumed to be related as shown here.

energy and buoyancy force, thus leading to a faster propagation. The faster propagation of the convective deflagration wave results in a larger M_{Ni56}. For example, a variation of the propagation speed by 15% in the W6 – W8 models results in M_{Ni56} values ranging between 0.5 and $0.7 M_\odot$ (Nomoto et al. 1984), which could explain the observations.

In the delayed detonation model, M_{Ni56} is predominantly determined by the deflagration-to-detonation-transition (DDT) density ρ_{DDT}, at which the initially subsonic deflagration turns into a supersonic detonation (Khokhlov 1991). As discussed in Umeda et al. (1999b), ρ_{DDT} could be very sensitive to $X(C)$, and a larger $X(C)$ is likely to result in a larger ρ_{DDT} and M_{Ni56}.

Here we postulate that M_{Ni56} and consequently brightness of a SN Ia increase as the progenitors' C/O ratio increases (and thus $M_{WD,0}$ decreases). As illustrated in Figure 8, the range of $M_{Ni56} \sim 0.5 - 0.8 M_\odot$ is the result of an $X(C)$ range $0.35 - 0.5$, which is the range of $X(C)$ values of our progenitor models. The $X(C) - M_{Ni56} - M_{WD,0}$ relation we adopt is still only a working hypothesis, which needs to be proved from studies of the turbulent flame during explosion (e.g., Niemeyer & Hillebrandt 1995).

3.3. *Metallicity and age effects*

3.3.1. *Metallicity effects on the minimum $M_{WD,0}$*

As mentioned in §2.5, M_w is the metallicity-dependent minimum $M_{WD,0}$ for a WD to become an SN Ia (*strong wind condition* in Fig. 8). The upper bound $M_{WD,0} \simeq 1.07 M_\odot$ is imposed by the condition that carbon should not ignite and is almost independent of metallicity. As shown in Figure 8, the range of $M_{CO} \simeq M_{WD,0}$ can be converted into a range of $X(C)$. From this we find the following metallicity dependence for $X(C)$:

(1) The upper bound of $X(C)$, which is determined by the lower limit on M_{CO} imposed by the metallicity-dependent conditions for a strong wind, e.g., $X(C) \lesssim 0.51$, 0.46 and 0.41, for Z=0.02, 0.01, and 0.004, respectively.

(2) On the other hand, the lower bound, $X(C) \simeq 0.35 - 0.33$, does not depend much on Z, since it is imposed by the maximum M_{CO}.

(3) Assuming the relation between M_{Ni56} and $X(C)$ given in Figure 8, our model predicts the absence of brighter SNe Ia in lower metallicity environment.

3.3.2. *Age effects on the minimum $M_{WD,0}$*

In our model, the age of the progenitor system also constrains the range of $X(C)$ in SNe Ia. In the SD scenario, the lifetime of the binary system is essentially the main-sequence lifetime of the companion star, which depends on its initial mass M_2. HKNU99 and HKN99 have obtained a constraint on M_2 by calculating the evolution of accreting WDs for a set of initial masses of the WD ($M_{WD,0} \simeq M_{CO}$) and of the companion (M_2), and the initial binary period (P_0). In order for the WD mass to reach M_{Ia}, the donor star should transfer enough material at the appropriate accretion rates. The donors of successful cases are divided into two categories: one is composed of slightly evolved main-sequence stars with $M_2 \sim 1.7 - 3.6 M_\odot$ (for Z=0.02), and the other of red-giant stars with $M_2 \sim 0.8 - 3.1 M_\odot$ (for Z=0.02) (Fig. 4).

If the progenitor system is older than 2 Gyr, it should be a system with a donor star of $M_2 < 1.7 M_\odot$ in the red-giant branch. Systems with $M_2 > 1.7 M_\odot$ become SNe Ia in a time shorter than 2 Gyr. Likewise, for a given age of the progenitor system, M_2 must be smaller than a limiting mass. This constraint on M_2 can be translated into the presence of a minimum M_{CO} for a given age, as follows: For a smaller M_2, i.e. for the older system, the total mass which can be transferred from the donor to the WD is smaller. In order for M_{WD} to reach M_{Ia}, therefore, the initial mass of the WD, $M_{WD,0} \simeq M_{CO}$, should be larger. This implies that the older system should have larger minimum M_{CO} as indicated in Figure 8. Using the $X(C)$-M_{CO} and M_{Ni56}-$X(C)$ relations (Fig. 8), we conclude that WDs in older progenitor systems have a smaller $X(C)$, and thus produce dimmer SNe Ia.

3.4. *Comparison with observations*

The first observational indication which can be compared with our model is the possible dependence of the SN brightness on the morphology of the host galaxies. Hamuy et al. (1996) found that the most luminous SNe Ia occur in spiral galaxies, while both spiral and elliptical galaxies are hosts to dimmer SNe Ia. Hence, the mean peak brightness is lower in elliptical than in spiral galaxies.

In our model, this property is simply understood as the effect of the different age of the companion. In spiral galaxies, star formation occurs continuously up to the present time. Hence, both WD+MS and WD+RG systems can produce SNe Ia. In elliptical galaxies, on the other hand, star formation has long ended, typically more than 10 Gyr ago. Hence, WD+MS systems can no longer produce SNe Ia. In Figure 9, we show the frequency of the expected SN I for a galaxy of mass $2 \times 10^{11} M_\odot$ for WD+MS and WD+RG systems separately as a function of M_{CO}. Here we use the results of HKN99 and HKNU99, and the $M_{CO} - X(C)$ and $M_{Ni56} - X(C)$ relations given in Figure 8. Since a WD with smaller M_{CO} is assumed to produce a brighter SN Ia (larger M_{Ni56}), our model predicts that dimmer SNe Ia occur both in spirals and in ellipticals, while brighter ones occur only in spirals. The mean brightness is smaller for ellipticals and the total SN Ia rate per unit luminosity is larger in spirals than in ellipticals. These properties are consistent with observations.

The second observational suggestion is the radial distribution of SNe Ia in galaxies. Wang et al. (1997) and Riess et al. (1998) found that the variation of the peak brightness for SNe Ia located in the outer regions in galaxies is smaller. This behavior can be understood as the effect of metallicity. As shown in Figure 8, even when the progenitor age is the same, the minimum M_{CO} is larger for a smaller metallicity because of the metallicity dependence of the WD winds. Therefore, our model predicts that the maximum brightness of SNe Ia decreases as metallicity decreases. Since the outer regions of galaxies are thought to have lower metallicities than the inner regions (Zaritsky et al. 1994; Kobayashi & Arimoto 1999), our model is consistent with observations. Wang et al. (1997) also claimed that SNe Ia may be deficient in the bulges of spiral galaxies. This can be explained by the age effect, because the bulge consists of old population stars.

3.5. *Diversity of high redshift supernovae*

We have suggested that $X(C)$ is the quantity very likely to cause the diversity in M_{Ni56} and thus in the brightness of SNe Ia. We have then shown that our model predicts that the brightness of SNe Ia depends on the environment, in a way which is qualitatively consistent with the observations. Further studies of the propagation of the turbulent flame and the DDT are necessary in order to actually prove that $X(C)$ is the key parameter.

Our model predicts that when the progenitors belong to an old population, or to a low metal environment, the number of very bright SNe Ia is small, so that the variation in brightness is also smaller. In spiral galaxies, the metallicity is significantly smaller at redshifts $z \gtrsim 1$, and thus both the mean brightness of SNe Ia and its range tend to be smaller. At $z \gtrsim 2$ SNe Ia would not occur in spirals at all because the metallicity is too low. In elliptical galaxies, on the other hand, the metallicity at redshifts $z \sim 1 - 3$ is not very different from the present value. However, the age of the galaxies at $z \simeq 1$ is only about 5 Gyr, so that the mean brightness of SNe Ia and its range tend to be larger at $z \gtrsim 1$ than in the present ellipticals because of the age effect.

We note that the variation of $X(C)$ is larger in metal-rich nearby spirals than in high redshift galaxies. Therefore, if $X(C)$ is the main parameter responsible for the diversity of SNe Ia, and if the light curve shape (LCS) method is confirmed by the nearby SNe Ia data, the LCS method can also be used to determine the absolute magnitude of high redshift SNe Ia.

3.6. *Possible evolutionary effects*

In the above subsections, we consider the metallicity effects only on the C/O ratio; this is just to shift the main-sequence mass - $M_{WD,0}$ relation, thus resulting in no important evolutionary effect. However, some other metallicity effects could give rise to evolution of SNe Ia between high and low redshifts (i.e., between low and high metallicities).

Here we point out just one possible metallicity effect on the carbon ignition density in the accreting WD. The ignition density is determined by the competition between the compressional heating due to accretion and the neutrino cooling. The neutrino emission is enhanced by the *local* Urca shell process of, e.g., $^{21}Ne-^{21}F$ pair (Paczyński 1973). (Note

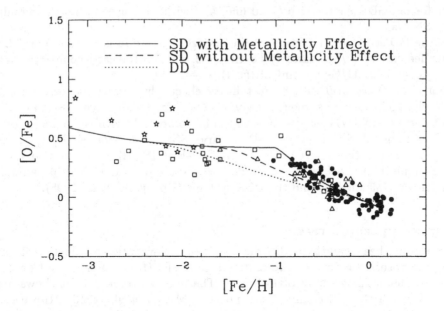

FIGURE 10. The evolutionary change in [O/Fe] against [Fe/H] for three SN Ia models. The dotted line is for the DD scenario, and the other lines are for our SD scenario with (solid line) and without (dashed line) the metallicity effect on SNe Ia. Observational data sources: filled circles, Edvardsson et al. (1993); open triangles, Barbuy & Erdelyi-Mendes (1989); stars, Nissen et al. (1994); open squares, Gratton (1991).

that this is different from the *convective* Urca neutrino process). For higher metallicity, the abundance of ^{21}Ne is larger so that the cooling is larger. This could delay the carbon ignition until a higher central density is reached (Nomoto et al. 1997d).

Since the WD with a higher central density has a larger binding energy, the kinetic energy of SNe Ia tends to be smaller if the same amount of ^{56}Ni is produced. This might cause a systematically slower light curve evolution at higher metallicity environment. The carbon ignition process including these metallicity effects as well as the convective Urca neutrino process need to be studied (see also Iwamoto et al. 1999 for nucleosynthesis constraints on the ignition density).

4. The chemical evolution in the solar neighborhood

The role of SNe II and SNe Ia in the chemical evolution of galaxies can be seen in the [O/Fe]-[Fe/H] relation (Fig. 10: Metal-poor stars with [Fe/H] $\lesssim$ −1 have [O/Fe] ∼ 0.45 on the average, while disk stars with [Fe/H] $\gtrsim$ −1 show a decrease in [O/Fe] with increasing metallicity. To explain such an evolutionary change in [O/Fe] against [Fe/H], we use the chemical evolution model that allows the infall of material from outside the disk region. The infall rate, the SFR, and the initial mass function (IMF) are given by Kobayashi et al. (1998), and the nucleosynthesis yields of SNe Ia and II are taken from Nomoto et al. (1997bc) and Tsujimoto et al. (1995).

The metallicity effects on SNe Ia on the chemical evolution is examined. For the DD scenario, the distribution function of the lifetime of SNe Ia by Tutukov & Yungelson (1994) is adopted, majority of which is ∼ 0.1−0.3 Gyr. Figure 10 shows the evolutionary change in [O/Fe] for three SN Ia models. The dotted line is for the DD scenario. The other lines are for our SD scenario with (solid line) and without (dashed line) the metallicity effect on SNe Ia.

(i) In the DD scenario the lifetime of the majority of SNe Ia is shorter than 0.3 Gyr. Then the decrease in [O/Fe] starts at [Fe/H] ~ -2, which is too early compared with the observed decrease in [O/Fe] starting at [Fe/H] ~ -1.

(ii) For the SD scenario with no metallicity effect, the companion star with $M \sim 2.6 M_\odot$ evolves off the main-sequence to give rise to SNe Ia at the age of ~ 0.6 Gyr. The resultant decrease in [O/Fe] starts too early to be compatible with the observations.

(iii) For the metallicity dependent SD scenario, SNe Ia occur at [Fe/H] $\gtrsim -1$, which naturally reproduce the observed break in [O/Fe] at [Fe/H] ~ -1. Also the low-metallicity inhibition of SN Ia provides a new interpretation of the SN II-like abundance patterns of the Galactic halo and the DLA systems (Kobayashi et al. 1998).

5. Cosmic supernova rates

Attempts have been made to predict the cosmic supernova rates as a function of redshift by using the observed cosmic star formation rate (SFR) (Ruiz-Lapuente & Canal 1998; Sadat et al. 1998; Yungelson & Livio 1998). The observed cosmic SFR shows a peak at $z \sim 1.4$ and a sharp decrease to the present (Madau et al. 1996). However, UV luminosities which is converted to the SFRs may be affected by the dust extinction (Pettini et al. 1998). Recent updates of the cosmic SFR suggest that a peak lies around $z \sim 3$.

Kobayashi et al. (1998) predicts that the cosmic SN Ia rate drops at $z \sim 1-2$, due to the metallicity-dependence of the SN Ia rate. Their finding that the occurrence of SNe Ia depends on the metallicity of the progenitor systems implies that the SN Ia rate strongly depends on the history of the star formation and metal-enrichment. The universe is composed of different morphological types of galaxies and therefore the cosmic SFR is a sum of the SFRs for different types of galaxies. As each morphological type has a unique star formation history, we should decompose the cosmic SFR into the SFR belonging to each type of galaxy and calculate the SN Ia rate for each type of galaxy.

Here we first construct the detailed evolution models for different type of galaxies which are compatible with the stringent observational constraints, and apply them to reproduce the cosmic SFR for two different environments, e.g., the cluster and the field. Secondly with the self-consistent galaxy models, we calculate the SN rate history for each type of galaxy and predict the cosmic supernova rates as a function of redshift.

5.1. *Supernova rates and galaxy types*

We assume that elliptical galaxies are formed by a single star burst and stop the star formation at $t \sim 1$ Gyr due to the supernova-driven galactic wind (e.g., Kodama & Arimoto 1997), while spiral galaxies are formed by a relatively continuous star formation. These models can well reproduce the present gas fractions, colors, supernova rates, and the color evolution in cluster ellipticals (see Figures 11 and 12).

Present supernova rates observed in the various type of galaxies (Cappellaro et al. 1997) put the constraints on the SN Ia progenitor models. Using the galaxy model shown in Figures 11 and 12, we show that our SN Ia model well reproduces the present supernova rates in both spirals and ellipticals (Kobayashi et al. 1999).

5.1.1. *Spiral galaxies*

The observed SN II rate R_{II} in late-type spirals is about twice the rate in early-type spirals. On the other hand, the observed SN Ia rate R_{Ia} in both types of spirals are

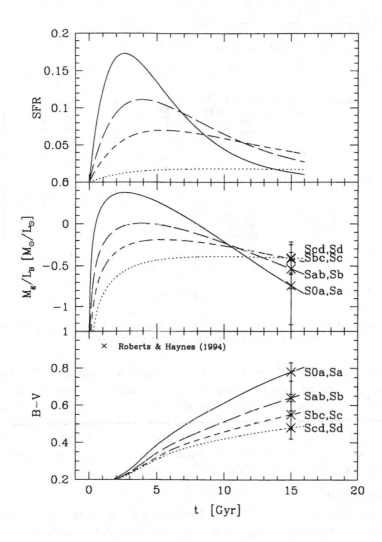

FIGURE 11. The star formation rates (top panel) gas fractions (middle panel) and the $B - V$ colors (bottom panel) in four types of spirals : S0a-Sa (solid line), Sab-Sb (long-dashes line), Sbc-Sc (short-dashed line), and Scd-Sd (dotted line). The present B-V colors are taken from Roberts & Haynes (1994), and the present gas fractions are normalized by the present blue luminosities and are derived from HI fractions (Roberts & Haynes 1994) and H_2/HI ratios (Casoli et al. 1998).

nearly the same. Therefore the present R_{Ia}/R_{II} ratio in early-type spirals is about twice that in late-type spirals. Such difference in the relative frequency might be a result of the difference in the star formation history, because the different dependences of R_{II} and R_{Ia} on the star formation history come from the different lifetimes of supernova progenitors. Therefore the observed $\mathcal{R}_{Ia}/\mathcal{R}_{II}$ ratio gives a constraint on the SN Ia progenitor model.

In the DD scenario, the lifetime of majority of SNe Ia is $\sim 0.1 - 0.3$ Gyr, so that the evolution of the SN Ia rate is similar to that of SN II rate. Therefore $\mathcal{R}_{Ia}/\mathcal{R}_{II}$ is insensitive to the star formation history. This results in the small differences among the various type of spirals, which is not consistent with observations.

In our SN Ia model, if iron abundance of progenitors are larger than [Fe/H]$\gtrsim -1$, the occurrence of SNe Ia is determined from the lifetime of the companions. If SNe Ia

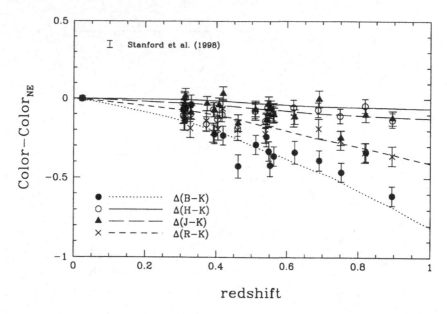

FIGURE 12. The passive color evolution of cluster ellipticals up to $z \sim 1$. The observational data are taken from Stanford et al. (1998).

would occur only in the MS+WD systems with relatively short lifetimes, $\mathcal{R}_{Ia}/\mathcal{R}_{II}$ would have been insensitive to the star formation histories in different types of spirals. On the contrary, if SNe Ia would occur only in the RG+WD systems with long lifetimes, the present difference in $\mathcal{R}_{Ia}/\mathcal{R}_{II}$ between early and late type spirals would have been too large, due to the large difference in SFR at an early epoch. Owing to the presence of these two types of the progenitor systems in our SN Ia progenitor model, the observed difference in $\mathcal{R}_{Ia}/\mathcal{R}_{II}$ can be reproduced.

5.1.2. *Elliptical galaxies*

We assume that a bulk of stars in cluster ellipticals are formed at $z \gtrsim 3$, and have ages $\gtrsim 10$ Gyr. Our SN Ia model, owing to the SN Ia lifetime of over 10 Gyr for WD+RG systems, well reproduces the present SNe Ia rate in ellipticals.

The longest lifetime of SNe Ia is ~ 11 Gyr for $Z = 0.002$ and ~ 18 Gyr for $Z = 0.02$; this is the lifetime of the smallest mass companion to produce SNe Ia, which is the $0.9M_\odot$ star, and its metallicity dependence. If the star formation in ellipticals has been stopped more than 10 Gyr before, majority of SNe Ia with $Z = 0.002$ are not found and only metal-rich SNe Ia occur at present. Accordingly, the SN Ia rate gradually decreases to the present, and the rate at $z \sim 0$ is smaller than that at $z \sim 0.2$ by $\sim 20\%$.

If ellipticals have undergone the relatively continuous star formation, as suggested by the observations of field ellipticals, the SN Ia rate keeps constant to the present. The tendency toward the present of the SN Ia rate in ellipticals can tell us the star formation histories in elliptical galaxies. Whether this tendency appears depends on whether the companions with a mass of $0.8M_\odot$ can produce SNe Ia, which should be examined.

5.2. *Cosmic supernova rates*

We calculate the cosmic star formation rate by summing up the SFRs of spirals and ellipticals with the ratio of the relative mass contribution among the types. The relative mass contribution is given by the relative luminosity proportion for ellipticals, S0a-Sa,

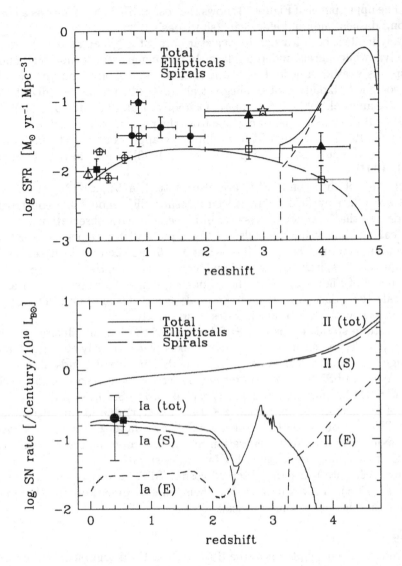

FIGURE 13. The upper panel shows the cosmic SFR (solid line) as a composite of spirals (long-dashed line) and ellipticals (short-dashed line). The dots are the observational data (Gallego et al. 1995, open triangle; Lilly et al. 1996, open circles; Madau et al. 1996, open squares; Connolly et al. 1997, filled circles; Tresse & Maddox 1998, open pentagon; Treyer et al. 1998, filled square; Glazebrook et al. 1999, filled pentagon; Hughes et al. 1998, star; Pettini et al. 1998, filled triangle). The lower panel shows the cosmic supernova rate (solid line) as a composite of spirals (long-dashed line) and ellipticals (short-dashed line). The observational data are from Pain et al. (1996) and Pain (1999).

Sab-Sb, Sbc-Sc, and Scd-Sd, respectively (Pence 1976), and the predicted mass to light ratio in B-band $(M/L_B)_i$ with our galaxy models (Figs. 11 and 12).

5.2.1. *In clusters*

First, we make a prediction of the cosmic supernova rates in the cluster galaxies using the galaxy models which are in good agreements with the observational constraints (Figs.

11 and 12). The upper panel of Figure 13 shows the cosmic SFR (solid line) as a composite of spirals (long-dashed line) and ellipticals (short-dashed line).

In our galaxy models, ellipticals undergo a star burst at $z \gtrsim 3$ and stop the star formation in ~ 1 Gyr, while spirals undergo relatively continuous star formation. Thus, only the SFR in spirals is responsible for the cosmic SFR at $z \lesssim 2$. Compared with the observational data, so-called Madau's plot (Gallego et al. 1995, filled triangle; Lilly et al. 1996, open circle; Madau et al. 1996, open square; Connolly et al. 1997, filled circle), the predicted cosmic SFR has a little shallower slope from the present to the peak at $z \sim 1.4$ (see also Totani et al. 1997). The high SFR in ellipticals appears at $z \gtrsim 3$. Such ellipticals may be hidden by the dust extinction (Pettini et al. 1998) or may have formed at $z \sim 5$ (Totani et al. 1997).

The recent data of the cosmic SFR are also plotted in Figure 13. However, these observations are taken by field galaxies. Our cosmic SFR model are constructed for cluster galaxies, so that it is not necessarily in agreement with observations.

The chemical enrichment in ellipticals has taken place with so much shorter timescale that the iron abundance reaches [Fe/H] ~ -1 at $z \sim 5$ and the metallicity effect on SNe Ia does not appear. In spirals, the iron abundance reaches [Fe/H] ~ -1 at $z \sim 2.6$.

The lower panel of Figure 13 shows the cosmic supernova rates (solid line) as a composite of spirals (long-dashed line) and ellipticals (short-dashed line). The upper and lower three lines show the SN II and Ia rates, respectively.

The SN Ia rate in spirals drops at $z \sim 2$ because of the low-metallicity inhibition of SNe Ia. In ellipticals, the chemical enrichment takes place so early that the metallicity is large enough to produce SNe Ia at $z \gtrsim 3$. The SN Ia rate depends almost only on the lifetime. A burst of SNe Ia occurs after ~ 0.5 Gyr which corresponds to the shortest lifetime of the WD+MS binaries; this forms a peak of the SN Ia rate at $z \sim 2.8$. The second peak of the SN Ia rate appears at $z \sim 1.8$ (~ 2 Gyr) due to the beginning of the explosions of SNe Ia from WD+RG progenitors. If SNe Ia at $z \gtrsim 2$ are observed with their host galaxies using the Next Generation Space Telescope, we can precisely test the metallicity effect by finding the drop of SN Ia rate in spirals.

The cosmic SN Ia rate using the observed cosmic SFR drops at $z \sim 1.6$ ($z \sim 1.2$ in Kobayashi et al. 1998). This is because the chemical enrichment of the whole universe is much slower than in spirals and ellipticals.

5.2.2. *In fields*

The observed spectra of ellipticals in the Hubble Deep Field suggest that the formation of ellipticals is protracted in fields, that is, the formation epochs of ellipticals span in the wide range of redshifts (Franceschini et al. 1998).

We also predict the cosmic supernova rates for the case that the formation of ellipticals in fields are protracted, that is, the formation epoch of ellipticals spans over the wide range of redshifts.

The upper panel of Figure 14 shows the cosmic SFR (solid line) as a composite of spirals (long-dashed line) and field ellipticals (short-dashed line).

The SFR in spirals is the same as in the upper panel of Figure 13, but the star formation in ellipticals continues to the present on the average. The peak of the star formation is at $z \sim 3$, which is consistent with the recent sub-mm data (Hughes et al. 1998). Then the synthesized cosmic SFR can successfully reproduce the observed one, except for the recent Hα data (Glazebrook et al. 1999).

The lower panel of Figure 14 shows the cosmic supernova rates (solid line) as a composite of spirals (long-dashed line) and ellipticals (short-dashed line). The upper and lower three lines show the SN II and Ia rates, respectively.

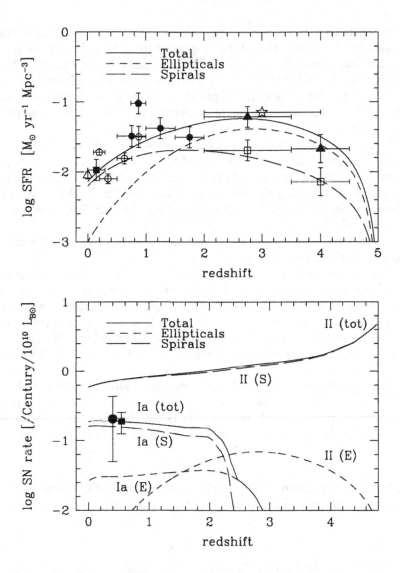

FIGURE 14. The same as Figure 13, but for the formation epochs of ellipticals spanning over $1 \lesssim z \lesssim 4$, which might correspond to field ellipticals.

The SN Ia rate in spirals drops at $z \sim 2$ as in the lower panel of Figure 13. Contrary to Figure 13, the SN Ia rate in field ellipticals gradually decreases from $z \sim 2$ to $z \sim 3$. This is because the star formation takes place more gradually on the average, and the number of stars becomes smaller toward higher redshifts. Since the chemical enrichment timescale in each elliptical is as fast as in the lower panel of Figure 13, the metallicity effect on SNe Ia is not effective. The decrease of the SN Ia rate in ellipticals is not due to the metallicity effect but due to the lifetime effect. We predict the different SN II and Ia rates between cluster and field ellipticals, which reflect the difference in the star formation histories in different environments.

5.2.3. *Summary*

We make a prediction of the cosmic supernova rate history as a composite of the different types of galaxies. We adopt the SN Ia progenitor scenario including the metallicity effect proposed by Kobayashi et al. (1998), which can successfully reproduce the chemical evolution of the solar neighborhood. To calculate the comic SFR, we construct the galaxy evolution models for spirals and ellipticals to meet the observational constraints such as the present gas fractions and colors for spirals, and the mean stellar metallicity and the color evolution from the present to $z \sim 1$ for ellipticals.

Owing to the *two* types of the progenitor system (MS+WD and RG+WD), i.e., shorter (0.6 − 1 Gyr) and longer lifetimes (2 − 15 Gyr) of SN Ia progenitors, we can explain the moderate contrast of the relative ratio of SN Ia to SN II rate $\mathcal{R}_{Ia}/\mathcal{R}_{II}$ between the early and late types of spirals. Owing to over 10 Gyr lifetime of the RG+WD systems, SNe Ia can be seen even at the present in ellipticals where the star formation has already stopped over 10 Gyr before.

Then we construct the cosmic SFR as the composite of the SFR for different types of galaxies, and predict the cosmic supernova rates:

(*a*) In the cluster environment, the synthesized cosmic SFR has an excess at $z \gtrsim 3$ corresponding to the SFR in ellipticals and a shallower slope from the present to the peak at $z \sim 1.4$, compared with Madau's plot. The predicted cosmic supernova rate suggests that SNe Ia can be observed even at high redshifts because the chemical enrichment takes place so early that the metallicity is large enough to produce SNe Ia at $z \gtrsim 3$ in cluster ellipticals. In spirals the SN Ia rate drops at $z \sim 2$ because of the low-metallicity inhibition of SNe Ia.

(*b*) In the field environment, ellipticals are assumed to form at such a wide range of redshifts as $1 \lesssim z \lesssim 4$. The synthesized cosmic SFR has a broad peak around $z \sim 3$, which well reproduces the observed one. The SN Ia rate is expected to be significantly low at $z \gtrsim 2$ because the SN Ia rate drops at $z \sim 2$ in spirals and gradually decreases from $z \sim 2$ in ellipticals.

This work has been supported in part by the grant-in-Aid for COE Scientific Research (07CE2002) of the Ministry of Education, Science, Culture and Sports in Japan.

REFERENCES

Arnett, W.D. 1996, Nucleosynthesis and Supernovae (Princeton: Princeton Univ. Press)

Barbuy, B., & Erdelyi-Mendes, M. 1989, A&A214 239

Branch, D. 1998, ARA&A, 1998, 36, 17

Branch, D., Lacy, C.H., McCall, M.L., Sutherland, P., Uomoto, A., Wheeler, J.C., & Wills, B.J. 1983, ApJ, 270, 123

Branch, D., Livio, M., Yungelson, L. R., Boffi, F. R., & Baron, E. 1995, PASP, 107, 717

Branch, D., Romanishin, W., & Baron, E., 1996, ApJ, 465, 73

Cappellaro, E., Turatto, M., Tsvetkov, D. Yu., Bartunov, O.S., Pollas, C., Evans, R., & Hammuy, M., 1997, A&A, 322, 431

Casoli, F., Sauty, S., Gerin, M., Boselli, A., Fouqué, P., Braine, J., Gavazzi, G., Lequeux, J., & Dickey, J. 1998, A&A, 331, 451

Connolly, A. J., Szalay, A. S., Dickinson, M., SubbaRao, M. U., & Brunner, R. J. 1997, ApJ, 486, L11

Cumming, R.J., Lundqvist, P., Smith, L.J., Pettini, M., & King, D.L. 1996, MNRAS, 283, 1355

Eck, C.R., Cowan, J.J., Roberts, D.A., Boffi, F.R., & Branch, D. 1995, ApJ, 451, L53

Edvardsson, B., Andersen, J., Gustafsson, B., Lambert, D. L., Nissen, P. E., & Tomkin, J. 1993, A&A, 275, 101

Franceschini, A., Silva, L., Fasano, G., Granato, G. L., Bressan, A., Arnouts, S. & Danese, L. 1998, ApJ, 506, 600

Gallego, J., Zamorand, J., Aragón-Salamanca, A., & Rego, M. 1995, ApJ, 455, L1

Garnavich, P. et al. , 1998, ApJ, 493, 53

Gilliland, R. L., Nugent, P. E., & Phillips, M. M. 1999, ApJ, in press (astro-ph/9903229)

Glazebrook, K., Blake, C., Economou, F., Lilliy S., & Colles, M. 1999, MNRAS, 306, 843

Gratton, R. G. 1991, in IAU Symp. 145, Evolution of Stars: The Photometric Abundance Connection, ed. G. Michaud & A. V. Tutukov (Montreal: Univ. Montreal), 27

Hachisu, I., & Kato, M., 1999, in preparation

Hachisu, I., Kato, M., & Nomoto, K., 1996, ApJ, 470, L97

Hachisu, I., Kato, M., & Nomoto, K. 1999, ApJ, 522, 487

Hachisu, I., Kato, M., Nomoto, K., & Umeda, H. 1999, ApJ, 519, 314

Hamuy, M., et al. 1995, AJ, 109, 1

Hamuy, M., Phillips, M. M., Schommer, R. A., & Suntzeff, N. B., 1996, AJ, 112, 2391.

Höflich, P., & Khokhlov, A., 1996, ApJ, 457, 500

Höflich, P., Nomoto, K., Umeda, H., & Wheeler, J. C., 1999, ApJ, submitted

Höflich, P., Wheeler, J. C., & Thielemann, F. -K., 1998, ApJ, 495, 617

Hughes, D., et al. 1998, Nature, 394, 241

Iben, I. Jr., & Tutukov, A. V. 1984, ApJS, 54, 335

Iglesias, C. A., & Rogers, F. 1993, ApJ, 412, 752

Iwamoto, K., Brachwitz, F., Nomoto, K., Kishimoto, N., Umeda, H., Hix, W. R., Thielemann, F-K. 1999, ApJS, in press

Kato, M., & Hachisu, I., 1999, ApJ, 513, L41

Khokhlov, A. 1991, A&A, 245, 114

Kobayashi, C., & Arimoto, N. 1999, ApJ, 526, in press

Kobayashi, C., Tsujimoto, T., & Nomoto, K. 1999, ApJ, submitted

Kobayashi, C., Tsujimoto, T., Nomoto, K., Hachisu, I, & Kato, M. 1998, ApJ, 503, L155

Kodama, T., & Arimoto, N., 1997, A&A, 320, 41

Kodama, T., Bower, R. G., & Bell, E. F. 1999, MNRAS, 306, 561

Li, X.-D., & van den Heuvel, E.P.J., 1997, A&A, 322, L9

Lilly, S. J., Le Fèvre, O., Hammer, F., & Crampton, D. 1995, ApJ, 460, L1

Madau, P., Ferguson, H. C., Dickinson, M. E., Giavalisco, M., Steidel, C. C., & Fruchter, A. 1996, MNRAS, 283, 1388

Mazzali, P. A., Cappellaro, E., Danziger, I. J., Turatto, M., & Benetti, S., 1998, ApJ, 499, L49

Niemeyer J.C., & Hillebrandt W. 1995, ApJ, 452, 769

Nissen, P. E., Gustafsson, B., Edvardsson, B., & Gilmore, G. 1994, A&A, 285, 440

Nomoto, K., 1982, ApJ, 253, 798

Nomoto, K., & Hashimoto, M. 1988, Phys. Rep., 163, 13

Nomoto, K., Iwamoto, K., & Kishimoto, N. 1997a, Science, 276, 1378

Nomoto, K., Hashimoto, M, Tsujimoto, T, Thielemann, F.-K, Kishimoto, N., Kubo, Y., & Nakasato, N. 1997b, Nuclear Physics, A616, 79c

Nomoto, K., Iwamoto, K., Nakasato, N., Thielemann, F.-K, Brachwitz, F., Tsujimoto, T., Kubo, Y., & Kishimoto, N. 1997c, Nuclear Physics, A621, 467c

Nomoto, K., et al. 1997d, in Thermonuclear Supernovae, Eds. P.Ruiz-Lapuente et al. (Dordrecht: Kluwer), 349

Nomoto, K., & Kondo, Y., 1991, ApJ, 367, l19

Nomoto, K., Nariai, K., & Sugimoto, D., 1979, PASJ, 31, 287

Nomoto, K., Thielemann, F. -K., & Yokoi, K., 1984, ApJ, 286, 644

Nomoto, K., Yamaoka, H., Shigeyama, T., Kumagai, S., & Tsujimoto, T. 1994, in Supernovae, Les Houches Session LIV, ed. S. A. Bludman et al. (Amsterdam: North-Holland), 199

Nugent, P., Baron, E., Branch, D., Fisher, A., & Hauschildt, P. H. 1997, ApJ, 485, 812

Paczyński, B. 1973, Acta Astr. 23, 1

Pain, R. 1999, Talk at the Type Ia supernova workshop in Aspen

Pain, R., et al. 1996, ApJ, 473, 356

Pei, Y. C., & Fall, S. M., 1995, ApJ, 454, 69

Pei, Y. C., Fall, S. M., & Hauser, M. G. 1999, ApJ, in press

Pence, W. 1976, ApJ, 420, L1

Perlmutter, S. et al. 1997, ApJ, 483, 565

Perlmutter, S. et al. 1999, ApJ, 517, 565

Pettini, M., Kellogg, M.,Steidel, C. C., Dickinson, M., Adelberger, K. L, & Giavalisco, M. 1998, ApJ, 508, 539

Phillips, M. M., 1993, ApJ, 413, L75

Riess, A.G., Press, W.H., & Kirshner, R.P. 1995, ApJ, 438, L17

Riess, A.G. et al. 1998, AJ, 116, 1009

Riess, A. G. et al. 1999, AJ, 117, 707

Roberts M. S. & Haynes M. P. 1994, ARAA, 32, 115

Ruiz-Lapuente, P., & Canal, R. 1998, ApJ, 497, L57

Sadat, R., Blanchard, A., Guiderdoni, B., & Silk, J. 1998, A&A, 331, L69

Saio, H., & Nomoto, K. 1985, A&A, 150, L21

Saio, H., & Nomoto, K. 1998, ApJ, 500, 388

Sandage, A., & Tammann, G.A., 1993, ApJ, 415, 1

Schlegel, E.M., & Petre, R. 1993, ApJ, 418, L53

Segretain, L., Chabrier, G., & Mochkovitch, R. 1997, ApJ, 481, 355

Stanford, S.A., Eisenhardt, P.R.M., & Dickinson, M., 1998, ApJ, 492, 461

Suzuki, T., & Nomoto, K. 1995, ApJ, 455, 658

Totani, T., Yoshii, Y., & Sato, K. 1997, ApJ, 483, L75

Tresse, L., & Maddox, S. J. 1998, ApJ, 495, 691

Treyer, M. A., Ellis, R. S., Milliard, B., Donas, J., & Bridges, T. J. 1998, MNRAS, 300, 303

Tsujimoto, T., Nomoto, K., Yoshii, Y., Hashimoto, M., Yanagida, S., & Thielemann, F.-K. 1995, MNRAS, 277, 945

Tutukov, A. V., & Yungelson, L. R. 1994, MNRAS, 268, 871

Umeda, H., Nomoto, K., Ymamaoka, H., & Wanajo, S. 1999a, ApJ, 513, 861

Umeda, H., Nomoto, K., Kobayashi, C., Hachisu, I, & Kato, M. 1999b, ApJL, 522, in press (astro-ph/9906192)

van den Heuvel, E.P.J., 1994, in Interacting Binaries, eds. S.N. Shore et al. (Berlin: Springer-Verlag), 263

van den Heuvel, E. P. J., Bhattacharya, D., Nomoto, K., & Rappaport, S. 1992, A&A, 262, 97

Wang, L., Höflich, P., & Wheeler, J. C. 1997, ApJ, 483, L29

Webbink, R. F. 1984, ApJ, 277, 355

Yungelson, L., & Livio, M. 1998, ApJ, 497, 168

Zaritsky, D., Kennicutt, R.C., & Huchra, J.P. 1994, ApJ, 420, 87

On the Evolution of Type Ia Supernovae with Redshift

By P. HÖFLICH[1], J. C. WHEELER[1], K. NOMOTP[2], H. UMEDA[2]

[1] Department of Astronomy, University of Texas, Austin, TX 78681, USA

[2] Department of Astronomy, Research Center for the Early Universe, Tokyo 113-0033, Japan

Based on detailed models for the explosions, light curves and NLTE-spectra, evolutionary effects of Type Ia with redshift have been studied to evaluate their size on cosmological time scales, how the effects can be recognized and how one may be able to correct for them.

In the first section, we summarize the current status of our models and put them into context with the empirical, statistical relations widely applied to use SNe Ia as yardsticks on cosmological distance scales. Both the statistical methods calibrated by δ-Ceph. stars and those based on theoretical model-fits for light curves (LC) and spectra of individual SNe Ia give consistent results for H_o. The statistical methods use the empirical relation between the absolute brightness and the postmaximum decline as a parameter to determine the absolute intrinsic brightness of an individual SNe Ia. From theoretical models, the amount of radioactive nickel produced during the explosion of a massive white dwarf has been identified as the basic quantity which provides this relation. The relation is independent from details of the model, however, the amount of ^{56}Ni actually produced depends on a combination of free parameters in the models such as central density and the chemical composition of the WD, and the propagation of the burning front. If these parameters are varied independently within the limits indicated by individual fits of 27 'classical' SNe Ia, we expect a spread around the mean relation of $\approx 0.4^m$ which is consistent with the spread based on the CTIO data published by Hamuy et al. (1995). Recent redefinitions of the statistical methods and new observations suggest a much tighter relation with a spread of $\approx 0.12^m$. This narrow spread cannot be understood in the context of current models. This tight relation may hint of an underlying coupling of the progenitor, the accretion rates and the propagation of the burning front. Here, we address some of the possible connections.

In the main section, we investigate the possible evolutionary effects in Type Ia supernovae. Detailed calculations have been performed for the explosion of a massive WD and the delayed detonation scenario which gives a good account of the optical and IR LCs and spectra.

We first address the influence of the initial metallicity, Z, and the C/O ratio on the nucleosynthesis, LCs and spectra. The C/O ratio of the typical progenitor is expected to change with redshift because it depends on the mass and, consequently, on the lifetime of the progenitor.

As the C/O ratio of the WD is decreased, the explosion energy and the ^{56}Ni production are reduced, and the Si-rich layers are more confined in velocity space. A reduction of C/O by about 60 % gives slower rise times by about 3 days, an increased luminosity at maximum light, a somewhat faster post-maximum decline and a larger ratio between maximum light and the ^{56}Co tail. With respect to the brightness decline relation, this translates into an offset of the zero point by about 0.3^m. A reduction of the C/O ratio has a similar effect on the colors, light curve shapes and element distribution as a reduction in the deflagration to detonation transition density but, for the same light curve shape, the absolute brightness is larger for smaller C/O. An independent determination of the initial C/O ratio and the transition density is possible for local SN if detailed analyses of both the spectra and light curves are performed simultaneously, or very early light curve observations are available.

Changing Z from Population I to II alters the isotopic composition of the outer layers of the ejecta that have undergone explosive O burning. Especially important is the increase of the ^{54}Fe production with Z. The influence on the resulting rest frame visual and blue light curves is found to be small. Although small, the changes in the spectrum with Z have an important effect on the colors of SN at high red shifts. The current high-z supernova searches use filters with the wavelength range corresponding to $\approx R$ and $\approx I$. Therefore, current searches are insensitive to the Z-effect up to about z=0.7 and 0.9, respectively. Systematic changes of up to 0.3^m can be expected at higher z.

In the second part, detailed stellar evolution calculations of a main sequence mass M_{MS} of 7 $M_\odot$ have been performed to test the influence of Z on the structure of an individual progenitor. A change of Z influences the central helium burning and, consequently, the size of the C/O core and the C/O ratio. Consequently, the C/O structure of the exploding white dwarf will depend on Z. As C and O are the fuel for the thermonuclear explosion, Z indirectly changes the energetics of the explosion. This governs the evolutionary effects to be expected for an individual SN Ia. In our example, changing Z from solar to 1/5 solar causes a change in the zero point of the maximum brightness/decline relation by about 0.1^m and a change in the rise time by about 1 day. Note that the exact size and sign of the effect of Z on the progenitor structure depends on the mass of the progenitor and it is not monotonic as a function of Z (Dominguez et al. 1999, Umeda et al. 1999). More systematic studies are required to provide quantitative tools to determine and correct evolutionary effects in detail.

Both the change of the 'typical' progenitor mass and of the influence of Z on the stellar evolution will effect the C/O ratio which changes the nuclear energy release during the explosion. The presence and sign of the effect will unreveal itself by the change in the rise time in combination with the expansion velocities of the various elements. The rise time may provide a direct measure of the potential size of the systematic effects. The offset in the maximum brightness/decline relation $\Delta M \sim \Delta t$ where Δt is the change of the rise time.

The evolutionary effects discussed here remain about a few tenths of a magnitude. Even strong variations in Z will hardly effect the counting rates for SNe Ia at large red-shifts. Variations of the expected size may prove to be absolutely critical if, in the future, SNe Ia are be used to measure large scale scalar-fields because, early on, Z may show large local variations.

In conclusion, evolutionary effects may be of the same order as the brightness changes as a function of cosmological parameters, but we have shown how the evolutionary effects can be detected.

---

1. Current Status

Two of the important developments in observational supernova research in the last few years were to establish the long-suspected correlation between the peak brightness of SNe Ia and their rate of decline by means of modern CCD photometry (Phillips 1993), and the exact distance calibrations provided by an HST Key Project (e.g. Saha et al. 1997). Both allowed a empirical determinations of H_o with unprecedented accuracy (Hamuy et al. 1996). Independent from these calibrations and empirical relations, H_o has been determined by comparisons of detailed theoretical models for light curves and spectra with observations (Müller & Höflich 1994, Höflich & Khokhlov 1996, Nugent et al. 1996). All determinations of the Hubble constant are in good agreement with one another. Recently, the routine successful detection of supernovae at large redshifts, z (Perlmutter et al. 1998, Riess et al. 1998), has provided an exciting new tool to probe cosmology. This work has provided results that are consistent with a low matter density in the Universe and, most intriguing of all, yielded hints for a positive cosmological constant. The cosmological results use empirical brightness-decline relations which are calibrated locally.

It is widely accepted that SNe Ia are thermonuclear explosions of carbon-oxygen white dwarfs. The primary scenario consists of massive carbon-oxygen white dwarfs (WDs) in close binary systems which accrete through Roche-lobe overflow from a low mass companion star when it evolves away from the main sequence or during its red giant phase (Nomoto & Sugimoto 1977). A WD is the final evolutionary state of stars with main

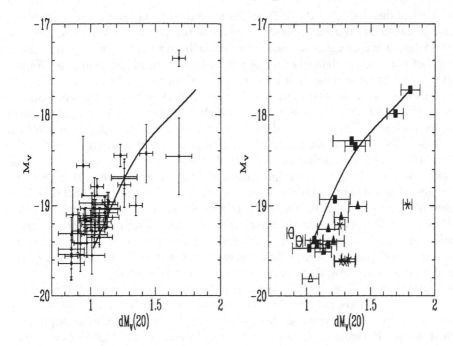

FIGURE 1. (Left panel) Observed light curve maximum brightness - decline rate relation. M_V is presented as a function of the decline from maximum at 20 days. (Right panel) The predicted relation for an array of models of SNe Ia representing delayed detonations (open triangles), pulsating delayed detonations (filled circles), merging models (open circles) and helium detonations (asterisks). For both the delayed detonation and merger scenarios models are only considered if they allow for a representation of some of the observed SNe Ia (Höflich et al. 1996).

sequence masses smaller than $\approx 7...8$ $M_\odot$ which has lost its H/He rich envelope with the C/O core of $\approx 0.5...1.2 M_\odot$. If accretion is sufficiently large (i.e. $\geq 2...4 \times 10^{-8} M_\odot/yr$), the accreted H burns to He and, subsequently, to C/O on the surface of the WD and the mass of the WD grows close to the Chandrasekhar limit. In these accretion models, the explosion is triggered by compressional heating close to the center.

Höflich, Khokhlov & Wheeler (1995) showed that models based on Chandrasekhar mass carbon-oxygen white dwarfs can account for subluminous as wells as "normally bright" SNe Ia. The basic paradigm of these models is that thermonuclear burning begins as a subsonic, turbulent deflagration and then makes a transition to a supersonic, shock-driven detonation (Khokhlov 1991ab). These models are generally known as delayed detonation models. Höflich et al. (1995) and Höflich & Khokhlov (1996) showed that in this class of models the amount of nickel produced is a function of the density at which the transition is made from deflagration to detonation †. The radioactive decay of the variable nickel mass gives a range in maximum brightness that matches the observations and identifies the amount of ^{56}Ni as the basic parameter which governs the light curve shape. The models with less nickel are not only dimmer, but are cooler and have lower opacity, giving them redder, more steeply declining light curves, in agreement with the observations (Fig. 1). From theory, the amount of ^{56}Ni produced during the explosion depends on a combination of parameters which, currently, are treated as free.

---

† We note that both the classical "deflagration" model W7 (Nomoto et al. 1984) and the "delayed detonation" models have a similar structure which has been successfully applied to reproduce light curves and spectra of normally bright SNe Ia (e.g. Harkness 1991).

They are the central density ρ_c of the WD at the time of the explosion, the density ρ_{tr} at which a transition to detonation occurs, the chemical profiles of the WD, and the progenitor scenario (since the merging scenario may contribute to the SNe Ia population). The realization of parameters depends on the coupled problem of progenitor evolution, accretion properties, flame propagation etc. which are not yet fully understood. For instances, ρ_c depends on the accretion rate, the progenitor structure, details of how the runaway occurs (Stein, private communication), which all depend on Z, the influence of Z on the star formation and so on. From detailed fits of 27 observed supernovae (Höflich & Khokhlov 1996), the range of variations of our free parameters cause a spread in the maximum brightness/decline relation of $\approx 0.4^m$. A similar spread of $\approx \pm 0.4^m$ in the local, observed maximum brightness decline/relation (Hamuy et al. 1995, Höflich et al. 1998) may suggest that there are additional parameters needed in the empirical relations. Note that new observations and recalibrations of the old observations indicate a much tighter relation ($\approx 0.12^m$, Fillipenko & Riess, this volume). From the current status of theoretical models, this very narrow spread cannot be understood, but it cannot be ruled out, either. Both analysis of new light curves and theoretical investigation of the coupling between the 'free' parameters may help to answer this puzzle.

There are already some hints that SN Ia have undergone evolutionary effects. Wang, Höflich, & Wheeler (1997) have presented evidence showing that SNe Ia are relatively underrepresented in the bulges of spiral galaxies and hence not selectively a bulge population as are classical novae. Perhaps even more intriguing, Branch et al. (1998) have shown that the mean peak brightness is dimmer in ellipticals than in spiral galaxies. Wang et al. (1998) found that the peak brightness in the outer region of spirals is similar to those found in ellipticals, but that in the central region both intrinsically brighter and dimmer SNe Ia occur. This implies that the progenitor populations are more inhomogeneous in the inner parts of spirals which contain both young and old progenitors.

Time evolution is expected to produce the following main effects: (a) a lower metallicity will decrease the time scale for stellar evolution of individual stars by about 20 % from Pop I to Pop II stars (Schaller et al. 1992) and, consequently, the progenitor population which contributes to the SNe Ia rate at any given time. The stellar radius also shrinks. This will influence the statistics of systems with mass overflow; (b) early on, we expect that systems with shorter life time will dominate the sample whereas, today, old system are contributing which may have not occurred early on. In addition, some scenarios with a life time comparable to the age of the univers such as two merging WDs may be absent a few Gyrs ago, but they may contribute today. (c) The initial metallicity will determine the electron to nucleon fraction of the outer layers and hence affects the products of nuclear burning; (d) Systems with a shorter life time may dominate early on and, consequently, the typical C/O ratio of the central region of the WD will be reduced; (e) The properties of the interstellar medium may change; (f) The influence of Z on the structure of WDs may change, but this effect has remains very small (e.g. 2% when comparing solar with 0.01 solar, Höflich et al. 1998). (g) The distribution of C and O will depend on Z as it influences the 'normal' stellar evolution and the properties of the C/O core. (h) The metallicity will effect nuclear burning during the accretion phase of the progenitor.

In the main part of this paper we address the possible influence of evolutionary effects on light curves and spectra for the delayed detonation scenario. In addition, the dependence of the WD core on the metallicity of the progenitor is studied for the example of a star with $M_{MS} = 7M_\odot$. Because the time evolution of the progenitor population is not well known and because we have not considered the entire variety of possible models, the values given below *do not* provide a basis for quantitative corrections of existing observations. The goal is to get a first order estimate of the size of the systematic effects and to demonstrate how evolutionary effects can be recognized in a real data sample and how one may be able to correct for them.

2. Brief Description of the Numerical Methods

2.1. *Stellar Evolution*

The stellar evolution has been calculated using the code of Nomoto's group up to the end of the helium burning. These calculations are discussed in detail by Umeda et al. (1999). Subsequently, the evolution of the C/O core is calculated by accreting H/He rich material at a given accretion rate on the core by solving the standard equations for stellar evolution using a Henyey scheme. Nomoto's equation of state is used (Nomoto et al. 1982). Crystallization is neglected. For the energy transport, conduction (Itoh et al. 1983), convection in the mixing length theory, and radiation is taken into account. Radiative opacities for free-free and bound free transitions are approximated in Kramer's approximation and by free electrons. A nuclear network of 35 species up to ^{24}Mg is used.

2.2. *Hydrodynamics*

The explosions are calculated using a one-dimensional radiation-hydro code, including nuclear networks (Höflich & Khokhlov 1996 and references therein). This code solves the hydrodynamical equations explicitly by the piecewise parabolic method (Collela & Woodward 1984) and includes the solution of the frequency averaged radiation transport implicitly via moment equations, expansion opacities (see below), and a detailed equation of state. Nuclear burning is taken into account using a network which has been tested in many explosive environments (see Thielemann et al. 1996 and references therein).

2.3. *Light Curves*

Based on the explosion models, the subsequent expansion and bolometric as well as monochromatic light curves are calculated using a scheme recently developed, tested and widely applied to SN Ia (e.g. Höflich et al. 1993 & 1998 and references therein) The code used in this phase is similar to that described above, but nuclear burning is neglected and γ ray transport is included via a Monte Carlo scheme. In order to allow for a more consistent treatment of scattering, we solve both the (two lowest) time-dependent, frequency averaged radiation moment equations for the radiation energy and the radiation flux, and a total energy equation. At each time step, we then use $T(r)$ to determine the Eddington factors and mean opacities by solving the frequency-dependent radiation transport equation in the comoving frame and integrate to obtain the frequency-averaged quantities. About one thousand frequencies (in one hundred frequency groups) and about five hundred depth points are used. The averaged opacities have been calculated under the assumption of local thermodynamic equilibrium. Both the monochromatic and mean opacities are calculated using the Sobolev approximation. The scattering, photon redistribution and thermalization terms used in the light curve opacity calculation are calibrated with NLTE calculations using the formalism of the equivalent-two-level approach (Höflich 1995).

2.4. *Spectral Calculations*

Our non-LTE code (Höflich et al. 1998 and references therein) solves the relativistic radiation transport equations in comoving frame. The energetics of the supernova are calculated. The evolution of the spectrum is not subject to any tuning or free parameters.

The non-LTE spectra are computed for various epochs using the chemical, density and

TABLE 1. Basic parameters for the delayed detonation models.

Model	C/O	R_Z	E_{kin}	M_{Ni}	Model	C/O	R_Z	E_{kin}	M_{Ni}
DD21c	1/1	1/1	1.32	0.69	DD25c	1/1	3/1	1.32	0.69
DD23c	2/3	1/1	1.18	0.59	DD26c	1/1	1/10	1.32	0.73
DD24c	1/1	1/3	1.32	0.70	DD27c	1/1	10/1	1.32	0.69

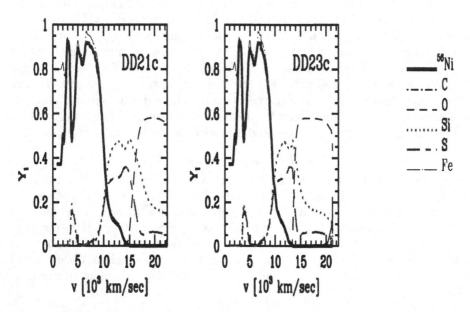

FIGURE 2. Abundances as a function of the final expansion velocity for the delayed detonation models DD21c and DD23c. Both the initial ^{56}Ni and the final Fe profiles are shown.

luminosity structure and γ-ray deposition resulting from the light curve code providing a tight coupling between the explosion model and the radiative transfer. The effects of instantaneous energy deposition by γ-rays, the stored energy (in the thermal bath and in ionization) and the energy loss due to the adiabatic expansion are taken into account. The radiation transport equations are solved consistently with the statistical equations and ionization due to γ radiation for the most important elements (C, O, Ne, Na, Mg, Si, S, Ca, Fe, Co, Ni). About 10^6 additional lines are included assuming LTE-level populations and an equivalent-two-level approach for the source functions.

3. Influence of the WD Structure on the Explosion

The influence of the initial metallicity and C/O ratio on light curves and spectra has been studied for the example of a set of delayed detonation models with the basic properties as follows: central density of the WD, $\rho_c = 2.6 \times 10^9 g/cm^{-3}$, $v_{burn} = \alpha * v_{sound}$ with $\alpha = 0.03$ during the deflagration phase and a transition to detonation at $\rho_{tr} = 2.7 \times 10^7$. The quantities of Table 1 in columns 2 to 5 and 6 to 9 are: C/O ratio; R_Z the Z relative to solar by mass; E_{kin} kinetic energy (in $10^{51} erg$); M_{Ni} mass of ^{56}Ni (in solar units). The parameters are close to those which reproduce both the spectra and light curves reasonably well (Nomoto et al. 1984; Höflich 1995).

3.1. *Influence of C/O*

As the C/O ratio of the WD is decreased from 1 to 2/3, the explosion energy and the ^{56}Ni production are reduced and the Si-rich layers are more confined in velocity space (Fig. 2). A reduction of C/O by about 60 % gives slower rise times by about 3 days and an increased luminosity at maximum light, a somewhat faster post-maximum decline and a larger ratio between maximum light and the ^{56}Co tail (Fig. 3). The increase in luminosity at maximum light is caused by the smaller expansion rate. Consequently, less energy stored early on is wasted in expansion work, but contributes to the luminosity. The smaller ^{56}Ni production causes the reduction of the luminosity later on.

A reduction of the C/O ratio has a similar effect on the colors, light curve shapes and element distribution as a reduction in the deflagration to detonation transition density but, for the same light curve shape, the absolute brightness is larger for smaller C/O. Moreover, the kinetic energy is reduced by about 10 % (Table 1) and, consequently, the expansion velocity derived by the Doppler shift in the spectra becomes smaller by about 5% . An independent determination of the initial C/O ratio and the transition density is possible for local SN if detailed analyses of both the spectra and light curves are performed simultaneously.

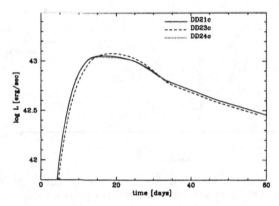

FIGURE 3. Comparison of bolometric light curves of the delayed detonation models DD21c, DD23c and DD24c with otherwise identical parameters but with different C/O ratios and Z relative to solar (C/O; R_Z) of (1;1), (2/3;1) and (1;0.3), respectively.

3.2. *Influence of the Metallicity*

To test the influence of the metallicity for nuclei beyond Ca, we have constructed models with parameters identical to DD21c but with initial metallicities between 0.1 and 10 times solar (Table 1). The energy release, the density and velocity structure are virtually identical to that of DD21c and, consequently, the bolometric (and also monochromatic) optical light curves are rather insensitive (Fig.3). The main influence of Z is a slight increase of the ^{56}Ni mass with decreasing Z due to a higher Y_e. The reason is that Z mainly effects the initial CNO abundances of a star. These are converted during the pre-explosion stellar evolution to ^{14}N in H-burning and via $^{14}N(\alpha,\gamma)^{18}F(\beta^+)^{18}O(\alpha,\gamma)^{22}Ne$ to nuclei with N=Z+2 in He-burning. The result is that increasing Z yields a smaller proton to nucleon ratio Y_e throughout the pre-explosive WD. Higher Z and smaller Y_e lead to the production of more neutron-rich Fe group nuclei and less ^{56}Ni (Fig. 4). For lower Z and, thus, higher Y_e, some additional ^{56}Ni is produced at the expense of ^{54}Fe and ^{58}Ni (Thielemann, Nomoto & Yokoi, 1986). The temperature in the inner layers is sufficiently high during the explosion that electron capture determines Y_e. In those layers, Z has no influence on the final burning product. The main differences due to changes in Z are in regions with expansion velocities in excess of $\approx$ 12000 km/sec. Most remarkable is the change in the ^{54}Fe production which is the dominant contributor to the abundance of iron group elements at these velocities since little cobalt has yet decayed near maximum light (Fig. 3).

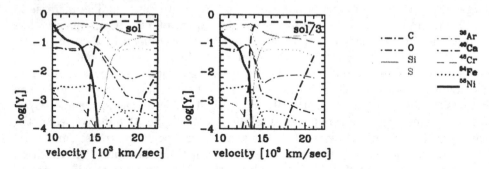

FIGURE 4. Abundances of different isotopes as a function of the expansion velocity for DD21c with initial abundances of solar and solar/3.

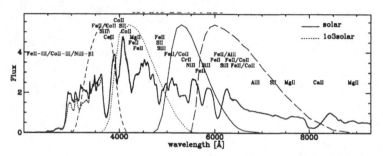

FIGURE 5. Comparison of synthetic NLTE spectra at maximum light for Z solar (DD21c) and 1/3 solar (DD24c). The standard Johnson filter functions for UBV, and R are also shown.

The initial WD composition has been found to have rather small effects on the overall LCs. The ^{56}Ni production and hence the bolometric and monochromatic optical and IR light curves differ only by a few hundredths of a magnitude. This change is almost entirely due to the small change in the ^{56}Ni production and not due to a change in the opacities because the diffusion time scales are governed by the deeper layers where burning is complete. However, the short wavelength part of the spectrum ($\lambda \leq 4000 \mathring{A}$) at maximum light is affected by a change in Z. This provides a direct test for Z of local SNe if well calibrated spectra are available and, thus, may give a powerful tool to unravel the nature (and lifetime) of SNe Ia progenitors.

By 2 to 3 weeks after maximum, the spectra are completely insensitive to the initial Z because the spectrum is formed in even deeper layers where none of the important abundances are effected by Z. Thus, for two similar bright SNe with similar expansion velocities, a comparison between the spectral evolution can provide a method to determine the difference in Z, or it may be used to detect evolutionary effects for distant SNe Ia if high quality spectra are available.

Although small, the spectral changes with Z have an important effect on the colors of SN at high red shifts where they are shifted into other bands. For local SNe Ia, a change of Z by a factor of 3 implies a variation in color of two to three hundredths of a magnitude. The high z-supernovae searches (Perlmutter et al. 1998; Riess et al. 1998) use filters with wavelengths in the wavelength range of the R and I bands. Therefore, these searches are insensitive to metallicity effects to about z=0.7 and 0.9, respectively. However, with the current filter sets, systematic changes of up to 0.3^m can be expected at larger redshifts (Höflich et al. 1998).

4. Influence of the Stellar Evolution on the WD Structure

Up to now, we have neglected the influence of the metallicity and the mass of the progenitor on the structure of the initial WD for a given mass on the main sequence. Such dependencies may become of important if supernovae are observed at large distance. On cosmological distance scales, Z is expected to be correlated with redshift. At the time of the explosion, the WD masses are close to the Chandrasekhar limit. The WD has grown by accretion of H/He and subsequent burning from the mass of the central core of a star with less than $\approx 7M_\odot$ (Fig. 6). In the accreted layers, the C/O ratio is close to 1; however, the initial mass of the C/O WD is given by the results of stellar evolution. The core mass depends on M_{MS} of the progenitor and on Z.

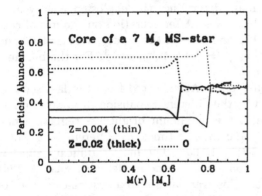

FIGURE 6. Final chemical profile of the C/O core of a star with $M_{MS} = 7M_\odot$.

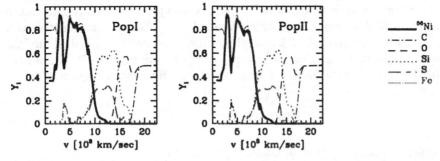

FIGURE 7. Abundances as a function of the final expansion velocity for progenitors of 7 $M_\odot$ with Z=0.02 and Z=0.004.

Here, we want to discuss the size of the metallicity effect using the example of a 7 $M_\odot$ model with Z=0.02 and 0.004. Z mainly effects the convection during the stellar Helium burning and, consequently, the size of the C/O core and the central C/O ratio. We note that the exact size of the effect and its sign depends on mass of the progenitor at the zero age main sequence, and Z. Even for a given mass, the changes are not monotonic, but may change sign from Pop I to Pop II to Pop III (Dominguez et al. 1999, Umeda et al. 1999). In addition, the tendency depends sensitively on the assumed physics such as the $^{12}C(\alpha,\gamma)^{16}O$-rate (e.g. Straniero et al. 1997). Therefore, our example can serve as a guide to estimate the size of this effect.

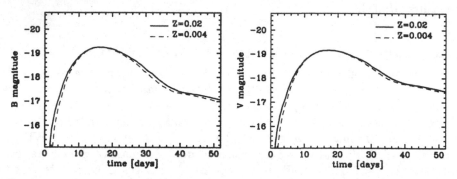

FIGURE 8. Comparison of light curves in B (left) and V (right) of the delayed de tonation models with metallicities with Z=0.02 and 0.004.

In agreement with §3.1, the total C/O mass determines the explosion energy and, consequently, has the main effect on the light curves. After accretion on the initial core M_C, the total mass fraction of M_C/M_{Ch} is 0.75 and 0.61 for the models with Z=0.04 and 0.02 respectively. At the time of the explosion, $\rho_c = 2.4 \times 10^9 g/cm^{-3}$. For the burning, $\alpha = 0.02$ and $\rho_{tr} = 2.4\ 10^7 g\ cm^{-3}$) (Fig. 7).

Monochromatic light curves are shown in Fig. 8. As expected from the last section, the main effect on the light curves is caused by the different expansion ratio determined by the integrated C/O ratio. The change in the maximum brightness remains small $(M_V(Z = 0.02) - M_V(Z = 0.04) = -0.02^m)$ and the rise times are different by about 1 day $(t_V((Z = 0.02) = 17.4d$ vs. $t_V(Z = 0.04) = 16.6d)$. The most significant effect is a steeper decline ratio and a reduced ^{56}Ni production for the model with solar Z. This is mainly due to the slower expansion ratio. This translates into a systematic offset of $\approx 0.1^m$ in the maximum brightness decline ratio (Hamuy et al. 1996). Using either the streching method or the LCS-method gives similar offsets.

Worth noting is the following trend: For realistic cores, the mean M_C/M_O tends to be smaller than the canonical value of 1 used in all calculations prior to 1998 (e.g. Nomoto et al. 1984, Woosley & Wheaver 1994, Höflich & Khokhlov 1996). Consequently, as a general trend, the rise times are about 1-2 days slower compared to the published models.

Acknowledgments: This work is supported in part by NASA Grant NAG5-7937 and NSF Grant AST-9528110, and a grant from the Texas Advanced Research Program.

REFERENCES

BRANCH D., et al. 1998, Phys. Rep., in press

COLLELA P., & WOODWARD P.R. 1984, J. Comp. Phys., 54, 174

DOMINGUEZ I., HÖFLICH P., STRANIERO O., WHEELER J.C. 1999, In Nucleosynthesis in the Cosmos (Kluver), 2

HAMUY M., et al. 1995, AJ, 109, 1

HAMUY M., et al. 1996, AJ, 112, 2438

HARKNESS R. P. 1991, In SN1987A and Other Supernovae, Eds Danziger & Kjär (ESO) 447

HÖFLICH P. 1995, ApJ, 443, 89

HÖFLICH P., KHOKHLOV A., WHEELER J. C., 1995, ApJ, 444, 831

HÖFLICH P., KHOKHLOV A., 1996, ApJ, 457, 500

HÖFLICH P., KHOKHLOV A., WHEELER J. C., PHILLIPS M., SUNTZEFF N., HAMUY M., 1996, ApJ, 472, 81

HÖFLICH P., WHEELER J. C., THIELEMANN F. K., 1997, ApJ, 495, 617

ITOH N., MITAKE S., IYETOMI H., ICHIMARU S., 1983, ApJ 273, 774

KHOKHLOV A., 1991ab , A&A, 245, 114 & L25

Müller E., Höflich P. 1994, *A&A* 281, 51

NOMOTO, K. & SUGIMOTO, D. 1977, PASJ, 29, 765

NOMOTO K., THIELEMANN F.-K., YOKOI K., 1984, ApJ, 286, 644

NOMOTO K., 1980, In IAU-Symp. 93, Ed. D. Reidel, 295

NUGENT P., et al. 1996, PRL, 75, 394 & 1974E

PHILLIPS M. M., 1993, ApJ, 413, L105

PHILLIPS M. M., et al. 1987, PASP, 90, 592

PERLMUTTER S., et al. 1998, ApJ, in press and astro-ph/9812473

RIESS A., et al. 1998, AJ 116, 1009

SAHA A., et al. 1997, ApJ, 486, 1

SCHALLER G., SCHAERER D., MEYNET G., MAEDER A., 1992, A&AS, 96, 269

THIELEMANN F.K., NOMOTO K., HASHIMOTO M., 1996, ApJ, 460, 408

THIELEMANN F.K., NOMOTO K., YOKOI K., 1986, A&A, 158, 17

UMEDA H., NOMOTO K., YAMAOKA H., WANAJO S., 1999, ApJ, 513, in press

WANG L., HÖFLICH P., WHEELER, J. C., 1997, ApJ, 487, L29

WOOSLEY S. E. & WEAVER T. A. 1994, In Proc. of Les Houches Session LIV, Eds. Bludman et al., (North-Holland) 63

Why Cosmologists Believe the Universe is Accelerating

By MICHAEL S. TURNER

Departments of Astronomy & Astrophysics and of Physics, Enrico Fermi Institute
The University of Chicago, Chicago, IL 60637-1433 USA
e-mail: mturner@oddjob.uchicago.edu

NASA/Fermilab Astrophysics Center, Fermi National Accelerator Laboratory
Batavia, IL 60510-0500 USA

Theoretical cosmologists were quick to be convinced by the evidence presented in 1998 for the accelerating Universe. I explain how this remarkable discovery was the missing piece in the grand cosmological puzzle. When found, it fit perfectly. For cosmologists, this added extra weight to the strong evidence of the SN Ia measurements themselves, making the result all the more believable.

1. Introduction

If theoretical cosmologists are the flyboys of astrophysics, they were flying on fumes in the 1990s. Since the early 1980s inflation and cold dark matter (CDM) have been the dominant theoretical ideas in cosmology. However, a key prediction of inflation, a flat Universe (i.e., $\Omega_0 \equiv \rho_{\text{total}}/\rho_{\text{crit}} = 1$), was beginning to look untenable. By the late 1990s it was becoming increasingly clear that matter only accounted for 30% to 40% of the critical density (see e.g., Turner 1999). Further, the $\Omega_M = 1$, COBE-normalized CDM model was not a very good fit to the data without some embellishment (15% or so of the dark matter in neutrinos, significant deviation from from scale invariance – called tilt – or a very low value for the Hubble constant; see e.g., Dodelson et al.1996).

Because of this and their strong belief in inflation, a number of inflationists (see e.g., Turner, Steigman & Krauss 1984 and Peebles 1984) were led to consider seriously the possibility that the missing 60% or so of the critical density exists in the form of vacuum energy (cosmological constant) or something even more interesting with similar properties (see Sec. 3 below). Since determinations of the matter density take advantage of its enhanced gravity when it clumps (in galaxies, clusters or superclusters), vacuum energy, which is by definition spatially smooth, would not have shown up in the matter inventory.

Not only did a cosmological constant solve the "Ω problem," but ΛCDM, the flat CDM model with $\Omega_M \sim 0.4$ and $\Omega_\Lambda \sim 0.6$, became the best fit universe model (Turner 1991 and 1997; Krauss & Turner 1995; Ostriker & Steinhardt 1995). In June 1996, at the Critical Dialogues in Cosmology Meeting at Princeton University, the only strike recorded against ΛCDM was the early SN Ia results of Perlmutter's group (Perlmutter et al.1997) which excluded $\Omega_\Lambda > 0.5$ with 95% confidence.

The first indirect experimental hint for something like a cosmological constant came in 1997. Measurements of the anisotropy of the cosmic background radiation (CBR) began to show evidence for the signature of a flat Universe, a peak in the multipole power spectrum at $l = 200$. Unless the estimates of the matter density were wildly wrong, this was evidence for a smooth, dark energy component. A universe with $\Omega_\Lambda \sim 0.6$ has a smoking gun signature: it is speeding up rather than slowing down. In 1998 came the SN Ia evidence that our Universe is speeding up; for some cosmologists this was a great surprise. For many theoretical cosmologists this was the missing piece of the grand puzzle and the confirmation of a prediction.

2. The theoretical case for accelerated expansion

The case for accelerated expansion that existed in January 1998 had three legs: growing evidence that $\Omega_M \sim 0.4$ and not 1; the inflationary prediction of a flat Universe and hints from CBR anisotropy that this was indeed true; and the failure of simple $\Omega_M = 1$ CDM model and the success of ΛCDM. The tension between measurements of the Hubble constant and age determinations for the oldest stars was also suggestive, though because of the uncertainties, not as compelling. Taken together, they foreshadowed the presence of a cosmological constant (or something similar) and the discovery of accelerated expansion.

To be more precise, Sandage's deceleration parameter is given by

$$q_0 \equiv \frac{(\ddot{R}/R)_0}{H_0^2} = \frac{1}{2}\Omega_0 + \frac{3}{2}\sum_i \Omega_i w_i \,, \tag{2.1}$$

where the pressure of component i, $p_i \equiv w_i \rho_i$; e.g., for baryons $w_i = 0$, for radiation $w_i = 1/3$, and for vacuum energy $w_X = -1$. For $\Omega_0 = 1$, $\Omega_M = 0.4$ and $w_X < -\frac{5}{9}$, the deceleration parameter is negative. The kind of dark component needed to pull cosmology together implies accelerated expansion.

2.1. *Matter/energy inventory:* $\Omega_0 = 1 \pm 0.2$, $\Omega_M = 0.4 \pm 0.1$

There is a growing consensus that the anisotropy of the CBR offers the best means of determining the curvature of the Universe and thereby Ω_0. This is because the method is intrinsically geometric – a standard ruler on the last-scattering surface – and involves straightforward physics at a simpler time (see e.g., Kamionkowski et al.1994). It works like this.

At last scattering baryonic matter (ions and electrons) was still tightly coupled to photons; as the baryons fell into the dark-matter potential wells the pressure of photons acted as a restoring force, and gravity-driven acoustic oscillations resulted. These oscillations can be decomposed into their Fourier modes; Fourier modes with $k \sim lH_0/2$ determine the multipole amplitudes a_{lm} of CBR anisotropy. Last scattering occurs over a short time, making the CBR is a snapshot of the Universe at $t_{ls} \sim 300,000$ yrs. Each mode is "seen" in a well defined phase of its oscillation. (For the density perturbations predicted by inflation, all modes the have same initial phase because all are growing-mode perturbations.) Modes caught at maximum compression or rarefaction lead to the largest temperature anisotropy; this results in a series of acoustic peaks beginning at $l \sim 200$ (see Fig. 1). The wavelength of the lowest frequency acoustic mode that has reached maximum compression, $\lambda_{\max} \sim v_s t_{ls}$, is the standard ruler on the last-scattering surface. Both $\lambda_{\max}$ and the distance to the last-scattering surface depend upon Ω_0, and the position of the first peak $l \simeq 200/\sqrt{\Omega_0}$. This relationship is insensitive to the composition of matter and energy in the Universe.

CBR anisotropy measurements, shown in Fig. 1, now cover three orders of magnitude in multipole and are from more than twenty experiments. COBE is the most precise and covers multipoles $l = 2 - 20$; the other measurements come from balloon-borne, Antarctica-based and ground-based experiments using both low-frequency ($f < 100\,\text{GHz}$) HEMT receivers and high-frequency ($f > 100\,\text{GHz}$) bolometers. Taken together, all the measurements are beginning to define the position of the first acoustic peak, at a value that is consistent with a flat Universe. Various analyses of the extant data have been carried out, indicating $\Omega_0 \sim 1 \pm 0.2$ (see e.g., Lineweaver 1998). It is certainly too early to draw definite conclusions or put too much weigh in the error estimate. However, a strong case is developing for a flat Universe and more data is on the way (Python V, Viper, MAT, Maxima, Boomerang, CBI, DASI, and others). Ultimately, the issue will

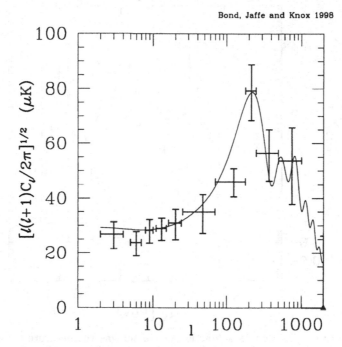

Bond, Jaffe and Knox 1998

FIGURE 1. Current CBR anisotropy data, averaged and binned to reduce error bars and visual confusion. The theoretical curve is for the ΛCDM model with $H_0 = 65\,\mathrm{km\,s^{-1}\,Mpc^{-1}}$ and $\Omega_M = 0.4$; note the goodness of fit (Figure courtesy of L. Knox).

be settled by NASA's MAP (launch late 2000) and ESA's Planck (launch 2007) satellites which will map the entire CBR sky with 30 times the resolution of COBE (around 0.1°).

Since the pioneering work of Fritz Zwicky and Vera Rubin, it has been known that there is far too little material in the form of stars (and related material) to hold galaxies and clusters together, and thus, that most of the matter in the Universe is dark. Weighing the dark matter has been the challenge. At present, I believe that clusters provide the most reliable means of estimating the total matter density.

Rich clusters are relatively rare objects – only about 1 in 10 galaxies is found in a rich cluster – which formed from density perturbations of (comoving) size around 10 Mpc. However, because they gather together material from such a large region of space, they can provide a "fair sample" of matter in the Universe. Using clusters as such, the precise BBN baryon density can be used to infer the total matter density (White et al.1993). (Baryons and dark matter need not be well mixed for this method to work provided that the baryonic and total mass are determined over a large enough portion of the cluster.)

Most of the baryons in clusters reside in the hot, x-ray emitting intracluster gas and not in the galaxies themselves, and so the problem essentially reduces to determining the gas-to-total mass ratio. The gas mass can be determined by two methods: 1) measuring the x-ray flux from the intracluster gas and 2) mapping the Sunyaev - Zel'dovich CBR distortion caused by CBR photons scattering off hot electrons in the intracluster gas. The total cluster mass can be determined three independent ways: 1) using the motions of clusters galaxies and the virial theorem; 2) assuming that the gas is in hydrostatic equilibrium and using it to infer the underlying mass distribution; and 3) mapping the cluster mass directly by gravitational lensing. Within their uncertainties, and where comparisons

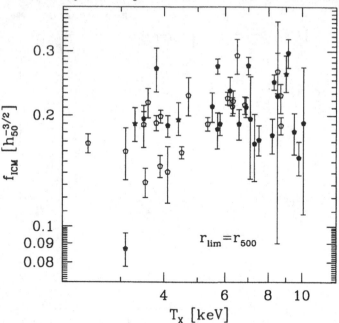

FIGURE 2. Cluster gas fraction as a function of cluster gas temperature for a sample of 45 galaxy clusters (Mohr et al., 1998). While there is some indication that the gas fraction decreases with temperature for $T < 5\,\mathrm{keV}$, perhaps because these lower-mass clusters lose some of their hot gas, the data indicate that the gas fraction reaches a plateau at high temperatures, $f_{\mathrm{gas}} = 0.212 \pm 0.006$ for $h = 0.5$ (Figure courtesy of Joe Mohr).

can be made, the three methods for determining the total mass agree; likewise, the two methods for determining the gas mass are consistent.

Mohr et al.(1998) have compiled the gas to total mass ratios determined from x-ray measurements for a sample of 45 clusters; they find $f_{\mathrm{gas}} = (0.075 \pm 0.002)h^{-3/2}$ (see Fig. 2). Carlstrom (1999), using his S-Z gas measurements and x-ray measurements for the total mass for 27 clusters, finds $f_{\mathrm{gas}} = (0.06 \pm 0.006)h^{-1}$. (The agreement of these two numbers means that clumping of the gas, which could lead to an overestimate of the gas fraction based upon the x-ray flux, is not a problem.) Invoking the "fair-sample assumption," the mean matter density in the Universe can be inferred:

$$\begin{aligned}
\Omega_M = \Omega_B/f_{\mathrm{gas}} &= (0.3 \pm 0.05)h^{-1/2} \text{ (Xray)} \\
&= (0.25 \pm 0.04)h^{-1} \text{ (S - Z)} \\
&= 0.4 \pm 0.1 \text{ (my summary)}.
\end{aligned} \qquad (2.2)$$

I believe this to be the most reliable and precise determination of the matter density. It involves few assumptions, most of which have now been tested. For example, the agreement of S-Z and x-ray gas masses implies that gas clumping is not significant; the agreement of x-ray and lensing estimates for the total mass implies that hydrostatic equilibrium is a good assumption; the gas fraction does not vary significantly with cluster mass.

2.2. Dark energy

The apparently contradictory results, $\Omega_0 = 1 \pm 0.2$ and $\Omega_M = 0.4 \pm 0.1$, can be reconciled by the presence of a dark-energy component that is nearly smoothly distributed. The cosmological constant is the simplest possibility and it has $p_X = -\rho_X$. There are other

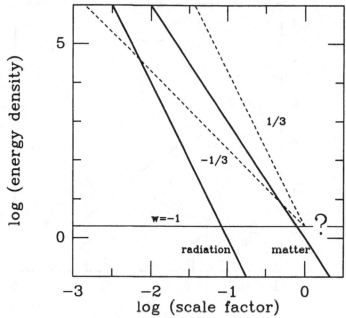

FIGURE 3. Evolution of the energy density in matter, radiation (heavy lines), and different possibilities for the dark-energy component ($w = -1, -\frac{1}{3}, \frac{1}{3}$) vs. scale factor. The matter-dominated era begins when the scale factor was $\sim 10^{-4}$ of its present size (off the figure) and ends when the dark-energy component begins to dominate, which depends upon the value of w: the more negative w is, the longer the matter-dominated era in which density perturbations can go into the large-scale structure seen today. These considerations require $w < -\frac{1}{3}$ (Turner & White 1997).

possibilities for the smooth, dark energy. As I now discuss, other constraints imply that such a component must have very negative pressure ($w_X \lesssim -\frac{1}{2}$) leading to the prediction of accelerated expansion.

To begin, parameterize the bulk equation of state of this unknown component: $w \equiv p_X/\rho_X$ (Turner & White 1997). This implies that its energy density evolves as $\rho_X \propto R^{-n}$ where $n = 3(1+w)$. The development of the structure observed today from density perturbations of the size inferred from measurements of the anisotropy of the CBR requires that the Universe be matter dominated from the epoch of matter – radiation equality until very recently. Thus, to avoid interfering with structure formation, the dark-energy component must be less important in the past than it is today. This implies that n must be less than 3 or $w < 0$; the more negative w is, the faster this component gets out of the way (see Fig. 3). More careful consideration of the growth of structure implies that w must be less than about $-\frac{1}{3}$ (Turner & White 1997).

Next, consider the constraint provided by the age of the Universe and the Hubble constant. Their product, $H_0 t_0$, depends the equation of state of the Universe; in particular, $H_0 t_0$ increases with decreasing w (see Fig. 4). To be definite, I will take $t_0 = 14 \pm 1.5\,\text{Gyr}$ and $H_0 = 65 \pm 5\,\text{km s}^{-1}\,\text{Mpc}^{-1}$ (see e.g., Chaboyer et al.1998); this implies that $H_0 t_0 = 0.93 \pm 0.13$. Fig. 4 shows that $w < -\frac{1}{2}$ is preferred by age/Hubble constant considerations.

To summarize, consistency between $\Omega_M \sim 0.4$ and $\Omega_0 \sim 1$ along with other cosmological considerations implies the existence of a dark-energy component with bulk pressure more negative than about $-\rho_X/2$. The simplest example of such is vacuum energy (Einstein's cosmological constant), for which $w = -1$. The smoking-gun signature of a

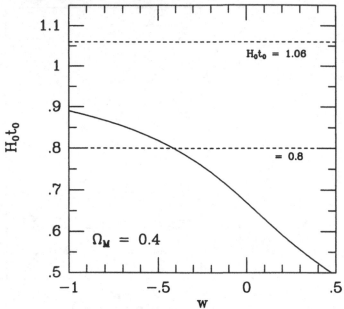

FIGURE 4. $H_0 t_0$ vs. the equation of state for the dark-energy component. As can be seen, an added benefit of a component with negative pressure is an older Universe for a given Hubble constant. The broken horizontal lines denote the 1σ range for $H_0 = 65 \pm 5\,\mathrm{km\,s^{-1}\,Mpc^{-1}}$ and $t_0 = 14 \pm 1.5\,\mathrm{Gyr}$, and indicate that $w < -\frac{1}{2}$ is preferred.

smooth, dark-energy component is accelerated expansion since $q_0 = 0.5 + 1.5 w_X \Omega_X \simeq 0.5 + 0.9w < 0$ for $w < -\frac{5}{9}$.

2.3. ΛCDM

The cold dark matter scenario for structure formation is the most quantitative and most successful model ever proposed. Two of its key features are inspired by inflation: almost scale invariant, adiabatic density perturbations with Gaussian statistical properties and a critical density Universe. The third, nonbaryonic dark matter is a logical consequence of the inflationary prediction of a flat universe and the BBN-determination of the baryon density at 5% of the critical density.

There is a very large body of data that is consistent with it: the formation epoch of galaxies and distribution of galaxy masses, galaxy correlation function and its evolution, abundance of clusters and its evolution, large-scale structure, and on and on. In the early 1980s attention was focused on a "standard CDM model": $\Omega_0 = \Omega_M = 1$, $\Omega_B = 0.05$, $h = 0.50$, and exactly scale invariant density perturbations (the cosmological equivalent of DOS 1.0). The detection of CBR anisotropy by COBE DMR in 1992 changed everything.

First and most importantly, the COBE DMR detection validated the gravitational instability picture for the growth of large-scale structure: The level of matter inhomogeneity implied at last scattering, after 14 billion years of gravitational amplification, was consistent with the structure seen in the Universe today. Second, the anisotropy, which was detected on the 10° angular scale, permitted an accurate normalization of the CDM power spectrum. For "standard cold dark matter", this meant that the level of inhomogeneity on all scales could be accurately predicted. It turned out to be about a factor of two too large on galactic scales. Not bad for an ab initio theory.

With the COBE detection came the realization that the quantity and quality of data that bear on CDM was increasing and that the theoretical predictions would have to match their precision. Almost overnight, CDM became a ten (or so) parameter theory. For astrophysicists, and especially cosmologists, this is daunting, as it may seem that a ten-parameter theory can be made to fit any set of observations. This is not the case when one has the quality and quantity of data that will soon be available.

In fact, the ten parameters of CDM + Inflation are an opportunity rather than a curse: Because the parameters depend upon the underlying inflationary model and fundamental aspects of the Universe, we have the very real possibility of learning much about the Universe and inflation. The ten parameters can be organized into two groups: cosmological and dark-matter (Dodelson et al.1996).

Cosmological Parameters

(a) h, the Hubble constant in units of $100 \, \mathrm{km \, s^{-1} \, Mpc^{-1}}$.

(b) $\Omega_B h^2$, the baryon density. Primeval deuterium measurements and together with the theory of BBN imply: $\Omega_B h^2 = 0.02 \pm 0.002$.

(c) n, the power-law index of the scalar density perturbations. CBR measurements indicate $n = 1.1 \pm 0.2$; $n = 1$ corresponds to scale-invariant density perturbations. Many inflationary models predict $n \simeq 0.95$; range of predictions runs from 0.7 to 1.2.

(d) $dn/d\ln k$, "running" of the scalar index with comoving scale ($k =$ wavenumber). Inflationary models predict a value of $\mathcal{O}(\pm 10^{-3})$ or smaller.

(e) S, the overall amplitude squared of density perturbations, quantified by their contribution to the variance of the CBR quadrupole anisotropy.

(f) T, the overall amplitude squared of gravity waves, quantified by their contribution to the variance of the CBR quadrupole anisotropy. Note, the COBE normalization determines $T + S$ (see below).

(g) n_T, the power-law index of the gravity wave spectrum. Scale-invariance corresponds to $n_T = 0$; for inflation, n_T is given by $-\frac{1}{7}\frac{T}{S}$.

Dark-matter Parameters

(a) Ω_ν, the fraction of critical density in neutrinos ($= \sum_i m_{\nu_i}/90h^2$). While the hot dark matter theory of structure formation is not viable, we now know that neutrinos contribute at least 0.3% of the critical density (Fukuda et al.1998).

(b) Ω_X and w_X, the fraction of critical density in a smooth dark-energy component and its equation of state. The simplest example is a cosmological constant ($w_X = -1$).

(c) g_*, the quantity that counts the number of ultra-relativistic degrees of freedom. The standard cosmology/standard model of particle physics predicts $g_* = 3.3626$. The amount of radiation controls when the Universe became matter dominated and thus affects the present spectrum of density inhomogeneity.

A useful way to organize the different CDM models is by their dark-matter content; within each CDM family, the cosmological parameters vary. One list of models is:

(a) sCDM (for simple): Only CDM and baryons; no additional radiation ($g_* = 3.36$). The original standard CDM is a member of this family ($h = 0.50$, $n = 1.00$, $\Omega_B = 0.05$), but is now ruled out (see Fig. 5).

(b) τCDM: This model has extra radiation, e.g., produced by the decay of an unstable massive tau neutrino (hence the name); here we take $g_* = 7.45$.

(c) νCDM (for neutrinos): This model has a dash of hot dark matter; here we take $\Omega_\nu = 0.2$ (about $5 \, \mathrm{eV}$ worth of neutrinos).

(d) ΛCDM (for cosmological constant) or more generally xCDM: This model has a smooth dark-energy component; here, we take $\Omega_X = \Omega_\Lambda = 0.6$.

Figure 5 summarizes the viability of these different CDM models, based upon CBR measurements and current determinations of the present power spectrum of inhomogeneity (derived from redshift surveys). sCDM is only viable for low values of the Hubble

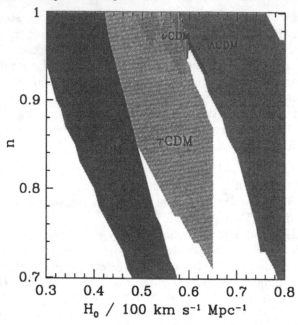

FIGURE 5. Summary of viable CDM models, based upon CBR anisotropy and determinations of the present power spectrum of inhomogeneity (Dodelson et al., 1996).

constant (less than $55\,\mathrm{km\,s^{-1}\,Mpc^{-1}}$) and/or significant tilt (deviation from scale invariance); the region of viability for τCDM is similar to sCDM, but shifted to larger values of the Hubble constant (as large as $65\,\mathrm{km\,s^{-1}\,Mpc^{-1}}$). νCDM has an island of viability around $H_0 \sim 60\,\mathrm{km\,s^{-1}\,Mpc^{-1}}$ and $n \sim 0.95$. ΛCDM can tolerate the largest values of the Hubble constant. While the COBE DMR detection ruled out "standard CDM," a host of attractive variants were still viable.

However, when other very relevant data are considered too – e.g., age of the Universe, determinations of the cluster baryon fraction, measurements of the Hubble constant, and limits to Ω_Λ – ΛCDM emerges as the hands-down-winner of "best-fit CDM model" (Krauss & Turner 1995; Ostriker & Steinhardt 1995; Turner 1997). At the time of the Critical Dialogues in Cosmology meeting in 1996, the only strike against ΛCDM was the absence of evidence for its smoking gun signature, accelerated expansion.

2.4. *Missing energy found!*

In 1998 evidence for the accelerated expansion anticipated by theorists was presented in the form of the magnitude – redshift (Hubble) diagram for fifty-some type Ia supernovae (SNe Ia) out to redshifts of nearly 1. Two groups, the Supernova Cosmology Project (Perlmutter et al.1998) and the High-z Supernova Search Team (Riess et al.1998), working independently and using different methods of analysis, each found evidence for accelerated expansion. Perlmutter et al. (1998) summarize their results as a constraint to a cosmological constant (see Fig. 7),

$$\Omega_\Lambda = \frac{4}{3}\Omega_M + \frac{1}{3} \pm \frac{1}{6}. \qquad (2.3)$$

For $\Omega_M \sim 0.4 \pm 0.1$, this implies $\Omega_\Lambda = 0.85 \pm 0.2$, or just what is needed to account for the missing energy! As I have tried to explain, cosmologists were quick than most to believe, as accelerated expansion was the missing piece of the puzzle found.

Recently, two other studies, one based upon the x-ray properties of rich clusters of

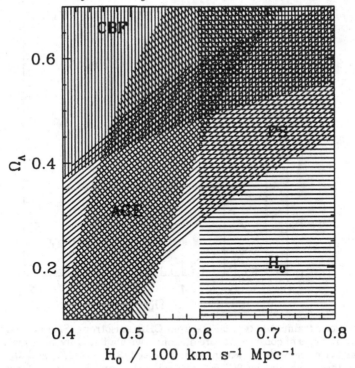

FIGURE 6. Constraints used to determine the best-fit CDM model: PS = large-scale structure + CBR anisotropy; AGE = age of the Universe; CBF = cluster-baryon fraction; and H_0= Hubble constant measurements. The best-fit model, indicated by the darkest region, has $H_0 \simeq 60 - 65\,\mathrm{km\,s^{-1}\,Mpc^{-1}}$ and $\Omega_\Lambda \simeq 0.55 - 0.65$. Evidence for its smoking-gun signature — accelerated expansion – was presented in 1998 (adapted from Krauss & Turner 1995 and Turner 1997).

galaxies (Mohr et al.1999) and the other based upon the properties of double-lobe radio galaxies (Guerra et al.1998), have reported evidence for a cosmological constant (or similar dark-energy component) that is consistent with the SN Ia results (i.e., $\Omega_\Lambda \sim 0.7$).

There is another test of an accelerating Universe whose results are more ambiguous. It is based upon the fact that the frequency of multiply lensed QSOs is expected to be significantly higher in an accelerating universe (Turner 1990). Kochanek (1996) has used gravitational lensing of QSOs to place a 95% cl upper limit, $\Omega_\Lambda < 0.66$; and Waga and Miceli (1998) have generalized it to a dark-energy component with negative pressure: $\Omega_X < 1.3 + 0.55w$ (95% cl), both results for a flat Universe. On the other hand, Chiba and Yoshii (1998) claim evidence for a cosmological constant, $\Omega_\Lambda = 0.7^{+0.1}_{-0.2}$, based upon the same data. From this I conclude: 1) Lensing excludes Ω_Λ larger than 0.8; 2) Because of the modeling uncertainties and lack of sensitivity for $\Omega_\Lambda < 0.55$, lensing has little power in strictly constraining Λ or a dark component; and 3) When larger objective surveys of gravitational-lensed QSOs are carried out (e.g., the Sloan Digital Sky Survey), there is the possibility of uncovering another smoking-gun for accelerated expansion.

2.5. *Cosmic concordance*

With the SN Ia results we have for the first time a complete and self-consistent accounting of mass and energy in the Universe. The consistency of the matter/energy accounting is illustrated in Fig. 7. Let me explain this exciting figure. The SN Ia results are sensitive to the acceleration (or deceleration) of the expansion and constrain the combination $\frac{4}{3}\Omega_M -$

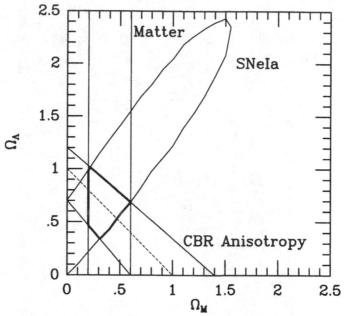

FIGURE 7. Two-σ constraints to Ω_M and Ω_Λ from CBR anisotropy, SNe Ia, and measurements of clustered matter. Lines of constant Ω_0 are diagonal, with a flat Universe shown by the broken line. The concordance region is shown in bold: $\Omega_M \sim 1/3$, $\Omega_\Lambda \sim 2/3$, and $\Omega_0 \sim 1$. (Particle physicists who rotate the figure by 90° will recognize the similarity to the convergence of the gauge coupling constants.)

Ω_Λ. (Note, $q_0 = \frac{1}{2}\Omega_M - \Omega_\Lambda$; $\frac{4}{3}\Omega_M - \Omega_\Lambda$ corresponds to the deceleration parameter at redshift $z \sim 0.4$, the median redshift of these samples). The (approximately) orthogonal combination, $\Omega_0 = \Omega_M + \Omega_\Lambda$ is constrained by CBR anisotropy. Together, they define a concordance region around $\Omega_0 \sim 1$, $\Omega_M \sim 1/3$, and $\Omega_\Lambda \sim 2/3$. The constraint to the matter density alone, $\Omega_M = 0.4 \pm 0.1$, provides a cross check, and it is consistent with these numbers. Further, these numbers point to ΛCDM (or something similar) as the cold dark matter model. Another body of observations already support this as the best fit model. Cosmic concordance indeed!

3. What is the dark energy?

I have often used the term exotic to refer to particle dark matter. That term will now have to be reserved for the dark energy that is causing the accelerated expansion of the Universe – by any standard, it is more exotic and more poorly understood. Here is what we do know: it contributes about 60% of the critical density; it has pressure more negative than about $-\rho/2$; and it does not clump (otherwise it would have contributed to estimates of the mass density). The simplest possibility is the energy associated with the virtual particles that populate the quantum vacuum; in this case $p = -\rho$ and the dark energy is absolutely spatially and temporally uniform.

This "simple" interpretation has its difficulties. Einstein "invented" the cosmological constant to make a static model of the Universe and then he discarded it; we now know that the concept is not optional. The cosmological constant corresponds to the energy associated with the vacuum. However, there is no sensible calculation of that energy (see e.g., Zel'dovich 1967; Bludman and Ruderman 1977; and Weinberg 1989), with estimates ranging from 10^{122} to 10^{55} times the critical density. Some particle physicists believe that when the problem is understood, the answer will be zero. Spurred in part by the possibility that cosmologists may have actually weighed the vacuum (!), particle theorists are taking a fresh look at the problem (see e.g., Harvey 1998; Sundrum 1997). Sundrum's proposal, that the gravitational energy of the vacuum is close to the present critical density because the graviton is a composite particle with size of order 1 cm, is indicative of the profound consequences that a cosmological constant has for fundamental physics.

Because of the theoretical problems mentioned above, as well as the checkered history of the cosmological constant, theorists have explored other possibilities for a smooth, component to the dark energy (see e.g., Turner & White 1997). Wilczek and I pointed out that even if the energy of the true vacuum is zero, as the Universe as cooled and went through a series of phase transitions, it could have become hung up in a metastable vacuum with nonzero vacuum energy (Turner & Wilczek 1982). In the context of string theory, where there are a very large number of energy-equivalent vacua, this becomes a more interesting possibility: perhaps the degeneracy of vacuum states is broken by very small effects, so small that we were not steered into the lowest energy vacuum during the earliest moments.

Vilenkin (1984) has suggested a tangled network of very light cosmic strings (also see, Spergel & Pen 1997) produced at the electroweak phase transition; networks of other frustrated defects (e.g., walls) are also possible. In general, the bulk equation-of-state of frustrated defects is characterized by $w = -N/3$ where N is the dimension of the defect ($N = 1$ for strings, $= 2$ for walls, etc.). The SN Ia data almost exclude strings, but still allow walls.

An alternative that has received a lot of attention is the idea of a "decaying cosmological constant", a termed coined by the Soviet cosmologist Matvei Petrovich Bronstein in 1933 (Bronstein 1933). (Bronstein was executed on Stalin's orders in 1938, presumably for reasons not directly related to the cosmological constant; see Kragh, 1996.) The term is, of course, an oxymoron; what people have in mind is making vacuum energy dynamical. The simplest realization is a dynamical, evolving scalar field. If it is spatially homogeneous, then its energy density and pressure are given by

$$\rho = \frac{1}{2}\dot{\phi}^2 + V(\phi)$$

$$p = \frac{1}{2}\dot{\phi}^2 - V(\phi) \tag{3.4}$$

and its equation of motion by (see e.g., Turner, 1983)

$$\ddot{\phi} + 3H\dot{\phi} + V'(\phi) = 0 \tag{3.5}$$

The basic idea is that energy of the true vacuum is zero, but not all fields have evolved to their state of minimum energy. This is qualitatively different from that of a metastable vacuum, which is a local minimum of the potential and is classically stable. Here, the field is classically unstable and is rolling toward its lowest energy state.

Two features of the "rolling-scalar-field scenario" are worth noting. First, the effective equation of state, $w = (\frac{1}{2}\dot{\phi}^2 - V)/(\frac{1}{2}\dot{\phi}^2 + V)$, can take on any value from 1 to -1. Second, w can vary with time. These are key features that may allow it to be distinguished from the other possibilities. The combination of SN Ia, CBR and large-scale structure data are already beginning to significantly constrain models (Perlmutter, Turner & White 1999), and interestingly enough, the cosmological constant is still the best fit (see Fig. 8).

The rolling scalar field scenario (aka mini-inflation or quintessence) has received a lot of attention over the past decade (Freese et al.1987; Ozer & Taha 1987; Ratra & Peebles 1988; Frieman et al.1995; Coble et al.1996; Turner & White 1997; Caldwell et al.1998).

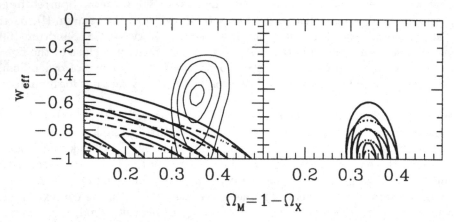

FIGURE 8. Contours of likelihood, from 0.5σ to 2σ, in the $\Omega_M - w_{\text{eff}}$ plane. Left: The thin solid lines are the constraints from LSS and the CMB. The heavy lines are the SN Ia constraints for constant w models (solid curves) and for a scalar-field model with an exponential potential (broken curves). Right: The likelihood contours from all of our cosmological constraints for constant w models (solid) and dynamical scalar-field models (broken). Note: at 95% cl w_{eff} must be less -0.6, and the cosmological constant is the most likely solution (from Perlmutter, Turner & White 1999).

It is an interesting idea, but not without its own difficulties. First, one must *assume* that the energy of the true vacuum state (ϕ at the minimum of its potential) is zero; i.e., it does not address the cosmological constant problem. Second, as Carroll (1998) has emphasized, the scalar field is very light and can mediate long-range forces. This places severe constraints on it. Finally, with the possible exception of one model (Frieman et al.1995), none of the scalar-field models address how ϕ fits into the grander scheme of things and why it is so light ($m \sim 10^{-33}$ eV).

4. Looking ahead

Theorists often require new results to pass Eddington's test: No experimental result should be believed until confirmed by theory. While provocative (as Eddington had apparently intended it to be), it embodies the wisdom of mature science. Results that bring down the entire conceptual framework are very rare indeed.

Both cosmologists and supernova theorists seem to use Eddington's test to some degree. It seems to me that the summary of the SN Ia part of the meeting goes like this: We don't know what SN Ia are; we don't know how they work; but we believe SN Ia are very good standardizeable candles. I think what they mean is they have a general framework for understanding a SN Ia, the thermonuclear detonation of a Chandrasekhar mass white dwarf, and have failed in their models to find a second (significant) parameter that is consistent with the data at hand. Cosmologists are persuaded that the Universe is accelerating both because of the SN Ia results and because this was the missing piece to a grander puzzle.

Not only have SN Ia led us to the acceleration of the Universe, but also I believe they will play a major role in unraveling the mystery of the dark energy. The reason is simple: we can be confident that the dark energy was an insignificant component in the past; it has just recently become important. While, the anisotropy of the CBR is indeed a cosmic Rosetta Stone, it is most sensitive to physics around the time of decoupling. (To

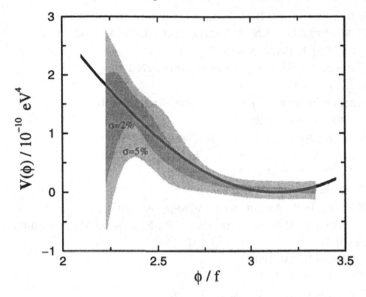

FIGURE 9. The 95% confidence interval for the reconstructed potential assuming luminosity distance errors of 5% and 2% (shaded areas) and the original potential (heavy line). For this reconstruction, $\Omega_M = 0.3$ and $V(\phi) = V_0[1 + \cos(\phi/f)]$ (from Huterer & Turner 1998).

be very specific, the CBR power spectrum is almost identical for all flat cosmological models with the same conformal age today.) SNe Ia probe the Universe just around the time dark energy was becoming dominant (redshifts of a few). My student Dragan Huterer and I (Huterer & Turner 1998) have been so bold as to suggest that with 500 or so SN Ia with redshifts between 0 and 1, one might be able to discriminate between the different possibilities and even reconstruct the scalar potential for the quintessence field (see Fig. 9).

My work is supported by the US Department of Energy and the US NASA through grants at Chicago and Fermilab.

REFERENCES

BRONSTEIN M. P., 1933, Phys. Zeit. der Sowjetunion, 3, 73

BLUDMAN S. & RUDERMAN M., 1977, PRL, 38, 255

CALDWELL R., DAVE R., & STEINHARDT P. J., 1998, PRL, 80, 1582

CARLSTROM, J., 1999, Physica Scripta, in press

CARROLL S., 1998, PRL, 81, 3067

CHABOYER B. et al., 1998, ApJ, 494, 96

CHIBA M. & YOSHII Y., 1998, ApJ, in press (astro-ph/9808321)

COBLE K., DODELSON S. & FRIEMAN J. A., 1996, PRD, 55, 1851

DODELSON S., GATES E. I. & TURNER M. S., 1996, Science, 274, 69

FREESE K. et al., 1987, Nucl. Phys. B, 287, 797

FRIEMAN J., HILL C., STEBBINS A. & WAGA I., 1995, PRL, 75, 2077

FUKUDA Y. ET AL., (SuperKamiokande Collaboration) 1998, PRL, 81, 1562

GUERRA E.J., DALY R.A. & WAN L., 1998, ApJ, submitted (astro-ph/9807249)

HARVEY J., 1998, hep-th/9807213.

HUTERER D. & TURNER M. S., 1998, PRL, submitted (astro-ph/9808133)

KAMIONKOWSKI M., SPERGEL D.N. & SUGIYAMA N., 1994, ApJ, 426, L57

KOCHANEK C., 1996, ApJ, 466, 638

KRAUSS L. & TURNER M. S., 1995, Gen. Rel. Grav., 27, 1137

LINEWEAVER C., 1998, ApJ, 505, L69

MOHR J., MATHIESEN B. & EVRARD A. E., 1998, ApJ, submitted.

MOHR J. et al., 1999, in preparation

OSTRIKER J. P. & STEINHARDT P. J., 1995, Nature 377, 600

OZER M. & TAHA M. O., 1987, Nucl. Phys. B, 287, 776

PEEBLES P. J. E., 1984, ApJ, 284, 439

PERLMUTTER S. et al., 1997, ApJ, 483, 565

PERLMUTTER S. et al., 1998, ApJ, in press (astro-ph/9812133)

PERLMUTTER S., TURNER M.S. & WHITE M., 1999, PRL, submitted (astro-ph/9901052)

RATRA B. & PEEBLES P. J. E., 1988, PRD, 37, 3406

RIESS A. et al., 1998, AJ, 116, 1009

SUNDRUM R., 1997, hep-th/9708329

SPERGEL D. N. & PEN U.-L., 1997, ApJ, 491, L67

TURNER E. L., 1990, ApJ, 365, L43

TURNER M. S., 1991, Physica Scripta, T36, 167

TURNER, M. S., 1997, In Critical Dialogues in Cosmology, Ed. N. Turok (World Scientific, Singapore) 555

TURNER M. S., 1999, Physica Scripta, in press (astro-ph/9901109)

TURNER M.S., STEIGMAN G. & KRAUSS L., 1984, PRL, 52, 2090

TURNER M.S. & WHITE M., 1997, PRD, 56, R4439

TURNER M.S. & WILCZEK F., 1982, Nature, 298, 633

VILENKIN A., 1984, PRL, 53, 1016

WAGA I. & MICELI A. P. M. R., 1998, astro-ph/9811460

WEINBERG S., 1989, Rev. Mod. Phys., 61, 1

WHITE S. D. M. et al., 1993, Nature 366, 429

ZEL'DOVICH YA. B., 1967, JETP, 6, 316

The Sloan Digital Sky Survey and Supernovae Ia

By Joshua A. Frieman (for the S D S S Collaboration)

NASA/Fermilab Astrophysics Center
Fermi National Accelerator Laboratory
P.O. Box 500, Batavia, IL 60510

Department of Astronomy & Astrophysics
The University of Chicago
5640 S. Ellis Avenue, Chicago, IL 60637

The Sloan Digital Sky Survey (SDSS) will produce a 3D map of the universe of galaxies over a volume $\sim 4 \times 10^7$ h^{-3} Mpc^3. Covering π sr. centered on the north galactic cap, the SDSS will comprise a photometric (CCD) imaging survey of 10^8 objects in 5 wavebands, a magnitude-limited spectroscopic (redshift) survey of 10^6 galaxies and 10^5 quasars, and a nearly volume-limited redshift survey of 10^5 bright red galaxies. The less-well-known southern SDSS will include repeated imaging of a ~ 225 sq. deg. region (to limiting magnitude $r' \sim 23$ per exposure) and should be useful for study of variable objects, including supernovae. This talk provides a brief overview of the SDSS and of its possible contributions to the study of Type Ia supernovae.

The Sloan Digital Sky Survey (SDSS) (York, etal. 1997) is a wide-area photometric and spectroscopic survey of the sky being carried out with a dedicated 2.5m telescope at Apache Point Observatory in southern New Mexico. While designed primarily to study galaxies, quasars, and their clustering properties in detail, the SDSS should prove a valuable tool for investigating a variety of variable phenomena in the sky, including supernovae. This short review provides an introduction to and status report on the SDSS, followed by some thoughts on detecting Type Ia SNe in the course of the survey.

1. Overview of the Sloan Survey

Fig. 1 shows a view of the Apache Point Observatory including the SDSS telescopes. While not large by present standards, the 2.5m telescope incorporates a number of novel design features. As the picture shows, the telescope is not housed in a conventional dome—its enclosure rolls away for observing, eliminating dome-induced temperature gradients and local turbulence which degrade image quality. In place of a dome, the rectangular structure shown mounted on the telescope provides wind and light baffling (it is mounted independently from the mirror support structure).

The digital imaging survey employs a large camera with an array of 30 primary CCD chips (each with 2048×2048 pixels of scale $0.4''$), arranged in 6 columns, with a corrected field of view $2.5°$ across. The chips in each column are covered by a sequence of 5 filters in the passbands u', g', r', i', z'. The photometry is carried out in long strips in drift-scan mode at the sidereal rate, with effective exposures of 55 sec, yielding an expected magnitude limit of $r' \simeq 23.1$ at $S/N = 5$ for point sources. Due to gaps between the CCD columns, two slightly offset scan strips are interleaved to make a contiguous $2.5°$ wide stripe. The 'main' Northern Survey will comprise 45 stripes to cover a contiguous area of π sr ($\sim 10,000$ sq. deg.) in the north galactic cap (see Fig. 2a), and will thus measure angular positions, magnitudes, and a large variety of image properties for about 100 million galaxies, a comparable number of stars, and about 10^6 QSOs. Nearly

FIGURE 1. Apache Point Observatory: the SDSS 2.5m telescope is shown at left, the structure in the center is the movable telescope enclosure, and the small dome at the right houses the Photometric Telescope, used for monitoring photometric standard stars. The large enclosure in the upper right is the ARC 3.5m telescope.

simultaneous imaging in multiple passbands is particularly useful for color-color selection of QSO candidates for follow-up spectroscopy and for estimating photometric redshifts of galaxies. The imaging will be carried out under the best observing conditions, i.e., on nights that are nearly 'photometric' and when the seeing is best—to cover the entire survey area requires ~ 20 % of the usable observing time over the five-year duration of the survey; the remainder will be devoted to spectroscopy. In order to tie the different imaging strips together into a uniform photometric system, a $20''$ photometric telescope simultaneously monitors photometric standard stars in the SDSS passbands.

The roughly 900,000 brightest galaxies ($r' < 18.1$), 10^5 color-selected bright red galaxies, 10^5 QSOs, and tens of thousands of stars, serendipitous objects, and ROSAT sources detected in the photometric survey will be targeted for follow-up spectroscopy. To accumulate redshifts at an unprecedented rate (of order 5000 per night), the SDSS will simultaneously use two 320-fiber spectrographs, each equipped with two CCD cameras, yielding spectra with resolution $\lambda/\Delta\lambda \simeq 2000$ over the wavelength range $3900 - 9200$ A. The 640 fibers for each spectroscopic exposure are plugged into pre-drilled holes in a round metal plate positioned in the focal plane of the telescope; the plate field of view is $3°$ across, and each fiber diameter subtends 3 arcsec on the sky. Each plate is exposed for 3×15 minutes.

In addition to the telescope and instrument hardware, the survey relies on a complex sequence of software to control the telescope and instruments during observing, and to carry out data reduction and processing; in addition, a number of tasks must be performed in coordinated fashion to prepare for each observing run. In advance of each photometric night, strategy software decides which strip to scan, given such variables as

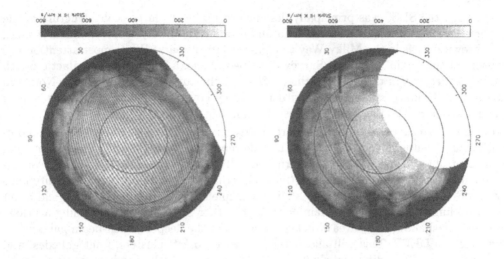

FIGURE 2. The footprints of the Northern and Southern SDSS. The curved lines show the survey stripes, which trace out great circles on the sky. The concentric circles indicate galactic latitudes $b = 0$, 30, and 60 deg. The grey scale map shows the Stark HI column density in units of 10^{20} cm^{-2}, which provides a visual impression of the expected extinction due to dust in our galaxy; the survey geometry is chosen to minimize these effects.

the time of year, weather patterns, and the strips previously completed. Once observed, an imaging strip corresponds to over 100 Gbytes of data, corresponding to ~ 10 Tbytes of raw data over the course of the survey. This data is processed through a photometric pipeline to extract detected objects and their properties. The resulting object catalogs are fed through the target selection pipeline which flags desired objects for spectroscopy. When a large enough contiguous area of sky has been imaged, and objects selected, an adaptive tiling routine decides how spectroscopic plug plates should be most efficiently laid down to cover that region. Once plates are designed, they are drilled and sent to the Observatory. The resulting spectra are run through a spectroscopic pipeline which automatically extracts spectra, measures redshifts, and outputs other spectroscopic information. The final photometric and spectroscopic data are archived in a science database designed for efficient data queries. After they are satisfactorily calibrated and carefully checked for systematics, the data products will be released to the public in annual or semi-annual stages during and after completion of the survey.

The telescope and imaging camera saw first light in May 1998 and have been undergoing commissioning. Since that time, imaging has been carried out over several dark runs, and several hundred square degrees of 5-band data have been taken along the celestial equator, with the telescope parked. In May 1999, the first spectrograph saw its first light on the telescope, and three extragalactic plates (each plugged with 320 fibers) were observed on June 8. Two of these plates were selected from SDSS imaging taken the previous March, successfully demonstrating 'end-to-end' operation of the survey. In the summer and Fall of 1999, the second spectrograph will be commissioned, and off-equatorial imaging drift scans will be performed. The Survey proper is expected to start in early 2000 and should take roughly 5 years to complete. For updated information about the survey and its status, see the SDSS Webpage at http://www.sdss.org.

2. Supernovae in the SDSS

Although the SDSS was principally designed to probe the large-scale structure of the universe, it will in fact provide a useful database for a wide variety of astronomical studies, from brown dwarfs in the Milky Way to high-redshift quasars. While most attention has focused on the 'main' Northern Survey described above, a survey in the galactic South will be carried out in the Fall months ($\sim$ 30 % of the survey time), when the Northern Survey area is inaccessible. The Southern Survey covers three 2.5° stripes at central declinations $\delta = 0°, -10°$, and $+15°$, as shown in Fig. 2b.

The Southern survey should prove particularly useful for the study of variable phenomena, including supernovae, since it will include repeated ($>$ 30 times) drift-scan imaging of the central (zero declination) southern stripe. This stripe covers $\sim 90°$ along the celestial equator; $6 + 6$ hours of scan time (for each strip of the stripe) produces a contiguous region of $\sim$ 225 square degrees. Co-adding multiple scans of this region will yield a net detection limit $\sim$ 2 magnitudes fainter than the other SDSS stripes, providing a bridge between the shallow, wide-area Northern survey and the deeper pencil-beam galaxy surveys, such as DEEP. This will allow study of high-redshift clusters, faint galaxies, and their evolution. In addition, differencing between scans can be used to search for variables. Examples of time-dependent phenomena that can be detected include Kuiper Belt Objects and stellar proper motions, variable QSOs and AGN, a variety of variable stars, and SNe.

Since the central Southern stripe lies along the celestial equator, it is observed with the telescope parked. This should improve the stability and astrometric accuracy of the imaging for this region compared to others in the Survey. How often will this region be observed? Perhaps 40 % of the dark observing time in the South will be devoted to imaging, i.e., roughly twice the fraction in the North. Although this means some compromise of image quality, that may be acceptable in obtaining light curves for certain variable objects. Given typical weather patterns, the fraction of dark time, etc, this leads to $\sim$ 540 hours of expected Southern imaging over 5 years, i.e., roughly 45 scans of the central southern stripe. These numbers are only meant to be suggestive of what can be accomplished. In fact, the Southern Survey observing strategy will likely evolve over the course of the SDSS: while the spectroscopy will be pushed somewhat deeper than in the North (to check the selection function and completeness of the Northern spectroscopic survey), one of the Southern observing seasons may be devoted predominantly to imaging (under admittedly mostly non-photometric conditions).

What contributions can SDSS make toward the study of SNe Ia? A back of the envelope estimate for the number of SNe Ia expected goes as follows (York, etal. 1997). Assume the apparent magnitude at peak luminosity of a Type Ia supernova in the blue is $m_B \simeq 18.8 + 5\log(z/0.1)$, with a variance of $\sigma \simeq 0.15$ mag after correction using the stretch factor, Δm_{15}, or multi-color light-curve shape. The estimated SNe Ia rate is $\Gamma_{Ia} \simeq 0.3h^2/(10^{10}L_\odot)$ per century. Using an estimated blue luminosity density of $\mathcal{L} \simeq 2 \times 10^8$ h $L_\odot$ Mpc^{-3} then yields roughly $7000(z/0.1)^3$ SNe Ia per year over the whole sky.

In the central Southern stripe, for each object we will have multi-color light curves with typically $\sim 8 - 9$ exposures spread (very probably non-uniformly) over ~ 3 months. With the numbers above, we will sample of order 20 SNe Ia per year at $z < 0.1$ with light-curves to 4 mag below peak and $\sim$ 200 SNe Ia per year at $z < 0.2$ with light curves to about 2 mag below peak. This assumes the photometry and image processing software perform well enough to subtract out their brighter host galaxies.

Thus one contribution of the SDSS to the study of SNe Ia will be to provide a uniformly selected sample of sparsely sampled multi-band light curves at relatively low redshift, $z < 0.2 - 0.3$. While this sample will not directly probe the presence of the cosmological constant or the issue of grey dust or evolution effects at high redshift, it will help fill in the current gap in the SNe Ia distribution at moderate redshifts. It could be useful for studying reddening and extinction effects at low redshift, the nature of SNe Ia in different galaxy populations, and could provide additional quantification of systematic errors.

It is also clear that the program above would be substantially enhanced by coordinating follow-up observations on other telescopes. This will be necessary for obtaining confirming spectra. Follow-up photometry on other facilities could also be useful to provide denser sampling of the SNe light-curves, as the mean sampling interval at the SDSS telescope may be $\sim 10 - 20$ days and is likely to be highly variable. If the fraction of Southern survey nights devoted to imaging is larger than 40 %, the latter issue will be less critical; however, one will ultimately be limited by weather at the site. On the other hand, a follow-up program places stringent constraints on the data reduction and processing timescale. Imaging data tapes are shipped to Fermilab for reduction through the photometric pipeline. Further analysis is then required to search for variable objects. In principle, this can be done within days, provided the computing system is not busy processing previous scans, but the survey is not generally set up to operate in this 'rapid response' mode. Other possibilities for rapid reduction/differencing of the imaging data have been under study by members of the collaboration. An advantage for follow-up is that these SNe are relatively bright at peak, so they can be followed with relatively modest telescopes. One might consider using the SDSS Photometric Telescope itself, if it is free of other tasks.

Given the large dataset the SDSS will produce, one might also consider exploring more speculative, non-traditional possibilities for detecting SNe beyond standard differencing techniques. For example, can one find SNe based on their colors? At very low redshift ($z \ll 0.1$), most SNe Ia should be deblended from their host galaxies by the photometric pipeline, given typical seeing conditions. One might therefore consider looking at regions of color space inhabited by SNe, i.e., the SDSS equivalent of $0 < B-V < 1, 0 < V-R < 0.5, -0.5 < R-I < 0.5$. One could contemplate selecting such objects and requiring the presence of a nearby bright galaxy with a plausible magnitude and photometric redshift. There are clearly many false positives which might overwhelm the selection, e.g., QSOs, A stars, white dwarfs, AGNs, etc. A second possibility is to discover SNe Ia spectroscopically. The SDSS will obtain 10^6 fiber spectra to 18th magnitude, with a mean galaxy redshift approaching $z = 0.15$. At that redshift, 10 kpc corresponds to an angular separation of 4.6h arcsec. Consequently, we expect perhaps 20 % of the SNe Ia to fall within the 3 arcsec fiber centered on a galaxy. At substantially lower redshifts, most will fall outside the fiber; at much higher redshifts, they will fall below the photometric detection threshold. However, in the intermediate regime, they could in principle be found by flagging cases where unusual broad lines are seen in the galaxy spectra.

I thank Scott Burles and Heidi Newberg for illuminating discussions and communications, particularly for informing me of the possibilities for non-standard SNe detection. Much of the rest has been gleaned from electronic communications of the SDSS Southern Working Group, in particular Michael Richmond, Craig Hogan, Jon Loveday, Bohdan Paczynski, Jeff Munn, and Heidi Newberg. This research was supported in part by the DOE and by NASA grant NAG5-7092 at Fermilab. The SDSS is a joint project of the University of Chicago, Fermilab, the Institute for Advanced Study, the Japan Partcipation Group, Johns Hopkins University, the Max-Planck-Institute for Astronomy, Princeton University, the United States Naval Observatory, and the University of

Washington. Apache Point Observatory, site of the SDSS, is operated by the Astrophysical Research Consortium. Funding for the project has been provided by the Alfred P. Sloan Foundation, the SDSS member institutions, the National Aeronautics and Space Administration, the National Science Foundation, the U.S. Department of Energy, and the Ministry of Education of Japan.

REFERENCES

For a complete description of the SDSS, see http://www.astro.princeton.edu/BBOOK/.

Type Ia supernovae and cosmology: Can $\Lambda = 0$ models be salvaged?

By WOLFGANG HILLEBRANDT

Max-Planck-Institut für Astrophysik, Karl-Schwarzschild-Str. 1, D-85740 Garching, Germany

From a theorist's biased point of view effects are discussed which might make type Ia supernovae (appear) systematically dimmer at high redshifts than their local counterparts and, thus, might provide loop-holes to make the observed supernova brightness - redshift relation compatible with $\Lambda = 0$ cosmologies. Such loop-holes could be hidden in fundamental uncertainties of supernova models, such as the still unknown nature of their progenitors, or the absence of a theory of turbulent thermonuclear combustion, but also in a variety of evolution effects which have not been studied carefully yet. From theory it cannot be excluded (and it even seems to be likely) that such effects could change type Ia light-curves in a systematic manner. However, they should show up also in sufficiently large nearby samples which, until now, is not observed. The reason could be that either the sample of well-studied local type Ia supernovae is still too small or they are indeed a more homogeneous class, as far as their observable properties are concerned, as theory would suggest.

1. Introduction

Systematic studies of type Ia supernovae at high redshifts between $z \simeq 0.3$ and 1 give increasing evidence that we are living in an expanding universe which began to *accelerate* its expansion when it was somewhat older than half its present age (Garnavich et al (1998a,b), Perlmutter et al (1997,1998); see also the contributions of Filippenko, Perlmutter, and others, in these proceedings). This finding is commonly interpreted as being due to a finite positive cosmological constant Λ (interpreted as the energy density of the vacuum) or, alternatively, attributed to a new form of yet unidentified energy density with negative pressure (Turner & White (1997), Caldwell et al (1998)).

Both explanations are not easy to swallow. First, if there is a positive cosmological constant: Why is it so small, but not zero? Also, it requires extreme fine-tuning of the initial conditions to have the cross-over between matter domination and vacuum domination at a redshift of about 0.5. As far as a new "dark" energy is concerned the situation is not much better. It would also be rather unpleasant if we would have to invent yet another unknown into cosmology in order to salvage the standard model. Moreover, the "dark" energy needs to have a very specific equation of state in order to be in agreement with the supernova data, but still requires extreme fine-tuning of the initial conditions. Finally, there are other arguments from observations in favor of a small (if not zero) cosmological constant. The number of gravitationally lensed quasars, which is proportional to the cosmic volume per unit redshift, seems to be incompatible with values of Λ in excess of what would give about $\frac{1}{2}$ of the critical density (Kochanek (1996), Falco et al (1998); see, however, Chiba & Yoshii (1998)). Also, the statistics of arcs seems to favor both, low matter density *and* small values of Λ (Bartelmann et al (1998)).

Therefore the question has to be addressed: Can we salvage $\Lambda = 0$ cosmologies in the light of the recent supervova observations? And, in fact, this workohop was intended to answer this question. In the following sections, a summary of some of the arguments, mainly outlining possible loop-holes in the interpretation of the supernova data, is presented. This summary is by no means complete but reflects to are large extend my own prejudices.

2. Possible problems with interpreting the observations

Roughly speaking, the effect we are talking about is that type Ia supernovae on average appear to be dimmer by about 0.25 mag as compared with their local relatives. There are certainly open questions related to the interpretation of these data. As seen by a theorist, they include some extra extinction not correlated with reddening in the same way as we are used to find it in our local neighborhood. For example, the properties of the interstellar dust in their host galaxies could be different, leading to some extinction *without* reddening. A systematic effect of that sort, although there is no evidence yet that it exists, could in fact make supernovae appear systematically dimmer at high redshifts.

A second possibility is that "peculiar" type Ia supernovae could be more frequent among the high redshift ones. Although, again, there is no indication that the distant sample is different from the local one, neither from the light curves nor from spectroscopy, such an effect cannot be excluded yet, mainly because both samples do not contain enough supernovae to allow to brake them up into several sub-samples and to carry out statistical tests.

Thirdly, there maybe evolutionary effects not obviously present in local supernovae which could have escaped detection. Metallicity, for example, can affect supernova light-curves and spectra in a variety of different ways, as will be discussed in more detail later. Of course, to first order, the same effects should show up also in nearby supernovae provided they span the same metallicity range. But given the still poor statistics of both samples some systematic discrepancy cannot be excluded from observations alone, leaving room for more theoretical investigations.

Next, the possibility that gravitational lensing might systematically de-magnify super-novae at high redshifts has also been looked at in the framework of simple models for the distribution of matter in the universe between us and the distant supernovae. However, it appears that this effect can account for at most $\Delta(\delta m) \leq 0.06$ mag, not enough to bring the cosmological constant back to zero (Holz (1998), Kolatt & Bartelmann (1998)).

A final problem concerns the light-curve-shape corrections (commonly called the *Phillips-relation* (Phillips (1993)) which have to be applied to the *local* sample in order to get a Hubble diagram with little scatter at *low* redshifts. It is often stated that the $\Lambda \neq 0$ conclusion is independent of these corrections, mainly because the photometric errors of the high-redshift observations are bigger than these corrections. However, as was pointed out by Leibundgut (1998a), the Phillips-relation (or any other one-parameter fit) fixes the "zero-point" of the brightness-redshift relation and thereby influences the best fit to Λ considerably. In fact, he finds that the light-curve corrections account for about one tenth of a magnitude. Without the corrections, $\Lambda = 0$ cosmologies could fit the supernova data rather well. On the other hand side, the reason why the peak luminosity - light-curve shape correlation works so well collapsing the nearby-supernova data into a well-defined Hubble-relation, is simply not known. In contrast to the observed relation, namely that brighter supernovae have a broader light-curve, one would expect naively that brighter supernovae are more energetic, expand faster and, therefore, also fade away faster. Therefore, if the correlation is correct and universal it must reflect effects coming from the transport of radiation in the expanding star. In that respect, it is a little disturbing that the correlation is not clearly present if UBVRI-light-curves are constructed (Leibundgut et al (1998), Contardo et al (1999)) although they should reflect the energetics more closely than the commonly used light-curves in the B-band.

All in all it appears reasonable, however, that the systematic errors are not big enough to bring Λ back to zero in an easy way (Schmidt et al 1998, Riess et al (1998), Perlmutter et al (1998)). The question therefore is: Can theory do this job, or are the cosmological parameters as determined from distant supernovae the final word and force us into searching for a some new physics to salvage the standard model of cosmology. This questions is addressed in the following sections.

3. Theoretical uncertainties

To rephrase the question given above: We have to find out, whether or not there exist loop-holes in the theoretical arguments through which models could predict systematically dimmer type Ia supernovae at high redshifts, preferentially by leaving the Phillips-relation intact. In general, we will concentrate the discussion on the most simple class of models, namely Chandrasekhar-mass white dwarfs, consisting of carbon and oxygen, and will mention alternative scenarios only in passing.

In principle, there are lots of problems with today's type Ia models. They start from the fact that we do not know their progenitors. A single white dwarf accreting matter from a normal companion is one option, but two white dwarfs merging are a possible alternative. We do not know for sure if at the time of the explosion the white dwarf has always reached the Chandrasekhar-mass or if, at least occasionally, a helium layer on top of a C+O white dwarf of lower mass ignites nuclear burning first, driving a burning front into the interior. Another fundamental problem is that we do not know the mode of thermonuclear burning by which the white dwarf is disrupted. It could be a sub-sonic deflagration wave, some kind of detonation, or a mixture of both.

Given all these uncertainties, it is impossible at present to offer models which are essentially parameter-free and make firm predictions as to what a type Ia should look like, as a function of environmental parameters. In contrast, all supernova models rely on a variety of assumptions and parameters which are then fitted to the observations. Since here we are concerned with the question if there could be systematic errors in interpreting the observed data this approach is, of course, dangerous, and one should rather try to find possible *discrepancies*. This is the approach I am going to use here. Moreover, since we are unable to settle any of the basic issues mentioned above I will concentrate on more technical and, thus, easier questions. Since obviously, systematic modifications in the appearance of type Ia supernovae could come from different metallicities, I will discuss possible (and until know not very carefully studied) effects in some detail.

3.1. *Accretion/merging process*

In what follows I shall always assume that the progenitors of type Ia supernovae are Chandrasekhar-mass C+O white dwarfs, leaving aside additional complications such as sub-Chandrasekhar models, having a layer of helium on top of a lower-mass C+O white dwarf. The reason is simply that the "typical" or "average" type Ia seems to be best fitted by the first class of models. I also will not discuss models in which two degenerate white dwarfs merge and form a critical mass for the ignition of carbon, mainly because binary systems which could lead to such a merger have not been identified so far. Moreover, the merging process will, in reality, be so complex that it is nearly impossible to construct realistic models. In any case, if white dwarf mergers should ever happen and give rise to type Ia supernovae, also for them metallicity effects could enter the picture, mainly because the distribution of white dwarf masses not only depends on the main sequence masses, but also on the metallicity Z directly (Umeda et al (1998)). Moreover, the C to O ratio is expected to depend on metallicity, as was recently shown by Umeda et al This, in turn, could have an effect on the mode on thermonuclear burning, as will be discussed later.

White dwarfs accreting mass from a companion star, on the other hand, are more likely candidates for type Ia progenitors, and the recently discovered super-soft x-ray sources are possible examples (see, e.g. , van den Heuvel et al (1992), Hashisu et al (1996), Li & van den Heuvel (1997), Yungelson & Livio (1998)). Here, metallicity can modify the progenitors in a variety of different ways, thereby introducing evolutionary effects of some sort which need to by looked at in detail.

The first effect is that, as before, the C/O-ratio in the interior of white dwarfs as well as the white dwarf/main sequence mass relationship changes with metallicity (Umeda et al (1998)). They found that, depending on the main sequence mass, the central C/O varies from 0.4 to 0.6, considerably less than assumed in most supernova models, but that for a given white dwarf mass metallicity effects on the C/O ratio as well as on the total amount of C of the final white dwarf are small, provided the accretion process itself is not changed.

This, however, needs not to be the case. More metals in the accreted gas can change, e.g., the accretion rate via the interaction with the radiation pressure or wind from the white dwarf (Kobayashi et al 1998)). Also, the limits on the accretion rate which allow for steady hydrogen burning without nova-like flashes is likely to depend on the CNO-abundances. In fact, it is possible that considerably lower accretion rates than commonly assumed (e.g. , Kahabka & van den Heuvel (1997)) can give rise to steady H-burning on top of rather massive white dwarfs (Kercek et al (1998)).

3.2. *URCA cooling or heating*

The thermal structure of a white dwarf on its way to an explosion is dependent on the (convective) URCA-process (Paczynski (1972), Iben (1978,1982), Barkat & Wheeler (1990), Mochkovitch (1996)). The URCA-pairs A = 21, 23, and 25 (such as, i.e., ^{21}Ne/^{21}F, ...) can lead to either heating or cooling, and possibly even to a temperature inversion near the center of the white dwarf. Naively one could expect that for a low carbon abundance also the abundances of the URCA-pairs should be low, leading to less cooling. Consequently, one would find a lower ignition density, less electron captures, more ^{56}Ni and, finally, a brighter light-curve. So the URCA-process could, in principle, make low metallicity supernovae even *brighter*! But this simple picture is certainly too naive.

3.3. *Evolution to ignition*

The evolution to ignition will be dependent again on the neutrino losses and β-decay energy release caused by the convective URCA-process, but now convection is definitely non-local, time-dependent, and 3-dimensional. Moreover, release of gravitational energy in the contracting white dwarf effects the thermal structure. It is not likely that in the near future we will be able to model all these processes in a realistic manner, even on super-computers, because of the vastly different time-scales involved. Again one might suspect that whatever happens should depend on, for example, the carbon abundance but it is hard to predict the consequences. Therefore, we have to rely on parameter studies, which is unpleasant, given the fact that we are interested in rather small effects.

3.4. *The ignition process*

Computer simulations of type Ia supernovae typically use a Chandrasekhar-mass white dwarf as initial condition, with a C/O-ratio of 1, and a temperature and density profile computed from quasi-static 1-dimensional stellar evolution codes ignoring, for example, the URCA-process discussed briefly in the previous sections. Since, in addition, the ignition itself is likely to be non-spherical, mainly because of buoyancy if the ignition is not exactly at the star's center (Garcia-Senz & Woosley (1995)), and since in the

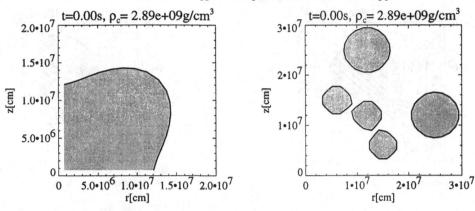

FIGURE 1. Initial ignition volumes of the models C(central)1 and B(blobs)5 of Reinecke et al (1998).

beginning most of the motions are still very sub-sonic, direct numerical simulations are not feasible. Therefore, also the ignition process has to be parameterized in some form. In principle, one could have ignition at or near the center in a small region, it could happen off-center, in particular if a temperature inversion should result from the URCA-process, it could happen in one, several, or many blobs, causally more or less disconnected. Which of these possibilities is realized could depend on evolution effects and, therefore, might modify typical ^{56}Ni-masses and, thus, light-curves.

Reinecke et al (1998) have recently performed a series of 2-dimensional simulations in which the mass, abundances, and thermal structure of the white dwarf was kept constant, but the places in the star where the nuclear fuel was ignited were different. The results were rather surprising (see Figs. 1 to 4): Changing the initial conditions from one central blob to several blobs off-center changed the final outcome from a white dwarf remaining bound to a (though week) explosion. The energy liberated by nuclear burning differed by 4×10^{50}erg! So, if for example low metallicity would lead to less URCA-cooling this could result in a low ignition density and *central* ignition, in contrast to a situation with high metallicity which could more often lead to off-center ignition with consequently higher Ni-masses and brighter light-curves. All this, of course, is speculation but indicates yet unexplored ways by which type Ia supernovae could, in principle, show the kind of evolutionary effects we are searching for.

3.5. *The physics of thermonuclear burning*

There is little doubt that at least in Chandrasekhar-mass models the burning front starts out as a sub-sonic deflagration wave. Because of the vastly different length-scales involved, ranging from fractions of millimeters for the width of the conductive nuclear flame at high densities to 1000 kilometers for the size of rising Rayleigh-Taylor blobs, direct numerical simulations are impossible, even in *one* dimension. Moreover, on macroscopic scales the burning front is propagated by turbulence rather than heat conduction, adding further complications to the picture.

Given all these problems, it is not surprising that up to now there is no agreement among the model-builders, neither quantitatively nor qualitatively, as to what the final outcome of a thermonuclear deflagration front in a Chandrasekhar-mass white dwarf is going to be. It could start out as a slow and largely sub-sonic deflagration. If it remains slow the star has time to expand before a large fraction of the C+O fuel is burned into nuclear statistical equilibrium. The result is a white dwarf still bound or a

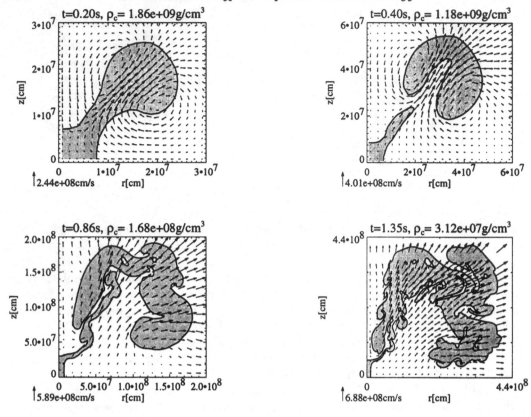

FIGURE 2. Time-sequence of simulation C1 of Reinecke et al (1998). Shown is the position of the burning front (as calculated by a level-set function) and the velocity field. At the top of each frame the time in seconds and the central density are given.

weak explosion, but nothing like a standard type Ia supernova (Niemeyer et al (1996), Reinecke et al (1998); see also Fig. 4). All successful models, such as W7 (Nomoto et al (1984)) or recent studies by Höflich et al (1995, 1998) parameterize the velocity of the deflagration front in order to fit observed properties of type Ia supernovae. For example, the turbulent flame speed is assumed to be a fraction of the local sound velocity, an assumptions that is very ad hoc and conflicts with numerical simulations which relate the turbulent velocity to the local shear created by hydrodynamic instabilities (Niemeyer & Hillebrandt (1995) Niemeyer et al (1996)). It is obvious that the "predictive power" of such parameterized models is not very big, and statements from theory concerning the presence or absence of evolution effects in supernova light-curves and spectra have to be taken with caution.

Best fits to light-curves an spectra are obtained from models which assume that at low densities, i.e. of the order of a few $10^7 \mathrm{gcm}^{-3}$, the deflagration front changes into a super-sonic detonation (Höflich & Khokhlov (1996)), driven by a pressure wave rather than by heat conduction or turbulence. One way how this might happen is thought to be a spontaneously transition when the front enters the so-called distributed burning regime where its thickness is no longer small compared with typical turbulent eddies and it gets heavily distorted by turbulence. The hope is that a sufficiently flat temperature gradient might form over a sufficiently large volume such that a critical mass can ignite nuclear burning almost instantaneously. However, a recent study making use

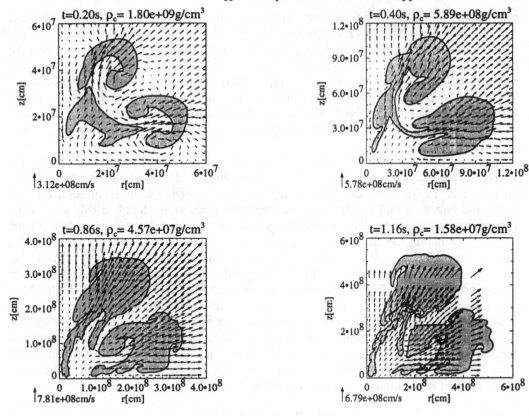

FIGURE 3. Same as Fig. 2, but for the sequence B5. The only obvious difference between the two simulations is a more structured burning front early in the evolution in B5, caused by the five blobs. This, however, changes the outcome even qualitatively, as can be seen in Fig. 4.

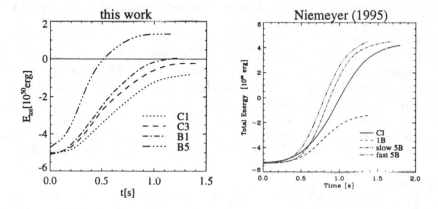

FIGURE 4. Total energies as a function of time for the various ignition conditions studied by Reinecke et al (1998). In addition to the models C1 and B5 discussed earlier the results of computations with three central "fingers" (C3) and one off-center blob (B1) are also given. For comparison, we present also results of Niemeyer et al (1996) who used a front capturing method but otherwise nearly identical initial conditions. Also in their calculations the explosion energy was strongly dependent on the structure of the volume that was incinerated as an initial condition.

of statistical properties of turbulence in the distributed regime seems to indicate that a spontaneous transition to a detonation is rather unlikely (Lisewski et al (1999)) and, in fact, under laboratory conditions it only seems to happen in containments or with burning fronts encountering obstacles. Moreover, under laboratory conditions, the transition to a detonation seems to require a quenching of the flame which can be obtained because of heat transport out of the containment or because of the particular kinetics of the chemical reactions involved. It is difficult to imagine similar physics in a star and, therefore, the prospects for prompt deflagration-detonation transitions in white dwarfs do not look promising.

A second possibility is that a partially burned but still bound white dwarf may re-contract and then go into a detonation. The model C1 discussed previously could be an example. In Fig. 2 it can be seen that towards the end of the simulation a lot of unburned matter is still located near the center. On the other hand, if this were to be the correct model for type Ia supernovae one should even be more surprised that they form such a homogeneous class of objects.

It is quite obvious that, whatever the explosion mechanism is, there are ways to introduce evolution effects. For example, it is easier to have a deflagration-detonation transition at low densities (if this ever happens) if the carbon abundance is high. Also, if the kind of turbulent nuclear burning, as outlined above, should be the cause of the explosion, additional parameters such as the amount and distribution of angular momentum would effect the outcome, simply because differential rotation is an extra source of turbulence. These latter parameters will, in turn, depend on the history of the white dwarf, its original mass, the mass accretion rate, etc. Again, it is difficult to see how all these degrees of freedom can lead to explosions being "standard candles".

3.6. *radiation transport*

I conclude this section with a few remarks on the problem of coupling the interior physics of an exploding white dwarf to what is finally observed, namely the physics of radiative transfer. For many astrophysical applications this problem is not solved, and type Ia supernovae are no exceptions. In fact, radiation transport is even more complex in type Ia's than for most other cases.

A rough sketch of the processes involved can illustrate some of the difficulties (see, e.g., Mazzali & Lucy (1993), Pinto & Eastman (1996)). Unlike most other objects we know in astrophysics type Ia supernovae do not contain any hydrogen. Therefore the opacities are always dominated by a huge number of atomic lines. In the beginning, the supernova is an opaque expanding sphere of matter into which energy is injected from radioactive decay. This could happen in a very inhomogeneous manner, as was discussed earlier. As the matter expands diffusion times eventually become shorter than the expansion time and the supernova becomes visual. However, because the star is rapidly expanding the Doppler-shift of atomic lines causes important effects. For example, a photon emitted somewhere in the supernova may find the surrounding matter more or less transparent until it finds a line Doppler-shifted such that it is trapped in that line and scatters many times. As a consequence, the spectrum might look thermal although the photon "temperature" has nothing in common with the matter temperature.

It is obvious that radiation transport in type Ia supernovae is very non-local and that the methods used commonly in models of stellar atmospheres need refinements. As a consequence, there is no agreement yet among the groups modeling light-curves and spectra as to what the best approach is (see the contributions of Höflich, Pinto, and Baron in these Proceedings). Therefore it can happen that even if the same model for the interior physics of the supernova is inserted into one of the existing codes for modeling light-curves and spectra the predictions for what should be "observed" could be different, again a very unpleasant situation. Things get even worth because all such models treat the exploding star as being spherically symmetric, an assumption that is at least questionable, given the complex combustion physics discussed earlier.

4. Conclusions

I have demonstrated that the physics of type Ia supernovae is very complex, starting from the possibility of very different progenitors, over the complexity of the physics leading to the explosion, to the complicated processes which couple the interior physics to observable quantities. None of these problems is fully understood yet, but what one can state is that it would appear to be a miracle if all the complexity would average out in a mysterious way to make type Ia supernovae easy-to-handle "standard candles" (or "standard bombs", as Alex Filippenko prefers to call them).

As it stands, the prediction from theory is that type Ia's should get more divers with increasing sample sizes. I would be very surprised if this prediction, which can only be tested with large samples of well-observed local supernovae, should turn out to be incorrect. I also would be surprised if this diversity could be absorbed in a "transformation" that contains just one parameter, such as the Phillips-relation or modifications of it.

This scepticism does not mean that I doubt that type Ia supernovae can be used to determine cosmological parameters. In contrast, I am sure that, in principle, they are excellent tools, possibly superior to most other methods. However, we need to understand them better before we can make full use of their potential. Unless we know for sure what kind of systematic errors need to be considered it could be dangerous to put too much weight on the numbers obtained from the assumption of (calibrated) standard candles. In that respect, it might be also too early to insist that there is new physics that needs to be included in cosmological models.

I would like to acknowledge helpful discussions with Jens Niemeyer and Stan Woosley, and the hospitality of University of California at Santa Cruz where this work was completed, supported in part by the DFG under grant Hi 534/3-1 and by the DAAD.

REFERENCES

BARKAT, Z., & WHEELER, J. C. 1990, ApJ, 355, 602

BARTELMANN, M., HUSS, A., COLBERG, J. M., JENKINS, A., & PEARCE, F. R. 1998, A&A 330, 1

CALDWELL, R. R., DAVE, R., & STEINHARDT, P. J. 1998, PRL, 80, 1582

CHIBA, M., & YOSHII, Y. 1998, ApJ, 489, 485

CONTARDO, G., LEIBUNDGUT, B., & VACCA, W. D. 1999, in preparation

FALCO, E. E., KOCHANEK, C. S., & MUNOZ, J. A. 1998, ApJ, 494, 47

GARCIA-SENZ, D., & WOOSLEY, S. E. 1995, ApJ, 454, 895

GARNAVICH, P. M., et al 1998, ApJ, 493, L53

GARNAVICH, P. M., et al 1998, ApJ, 509, 74

HASHISO, K., KATO, M., & NOMOTO, K. 1996, ApJ, 470, L97

HÖFLICH, P., & KHOKHLOV, A. 1996, ApJ, 457, 500

HÖFLICH, P., KHOKHLOV, A., & WHEELER, J. C. 1995, ApJ, 444, 211

HÖFLICH, P., WHEELER, J. C., & THIELEMANN, F. K. 1998, ApJ, 495, 617

HOLZ, D. E. 1998, ApJ, 506, L1

IBEN, I., JR. 1978, ApJ, 219, 213

IBEN, I., JR. 1982, ApJ, 253, 248

KAHABKA, P., & VAN DEN HEUVEL, E. P. J. 1997, ARAA, 35, 69

KERCEK, A., HILLEBRANDT, W., & TRURAN, J. W. 1998, A&A, in press

KOBAYASHI, C., TSUJIMOTO, T., NOMOTO, K., HASHISU, I., & KATO, M. 1998, ApJ, 503, L155

KOCHANEK, C. S. 1996, ApJ, 466, 638

KOLATT, T. S., & BARTELMANN, H. 1998, MNRAS, 296, 763

LEIBUNDGUT, B. 1998, in Supernovae and Cosmology, Eds. L. Labhardt, B. Binggeli, & R. Buser (U. Basel) 61

LEIBUNDGUT, B., CONTARDO, G., WOUDT, P.,& SPYROMILIO, J. 1998, astro-ph/9812042

LI, X.-D., & VAN DEN HEUVEL, E. P. J. 1997, A&A, 322, L9

LISEWSKI, M., HILLEBRANDT, W., WOOSLEY, S. E., NIEMEYER, J. C. & KERSTEIN, A. 1999, in preparation

MAZZALI, P. A., & LUCY, L. B. 1993, A&A, 279, 447

MOCHKOVITCH, R. 1996, A&A, 311, 152

NIEMEYER, J. C., & HILLEBRANDT, W. 1995, ApJ, 452, 769

NIEMEYER, J. C., HILLEBRANDT, W., & WOOSLEY, S. E. 1996, ApJ, 471, 903

NOMOTO, K., THIELEMANN, F. K., & YOKOI, K. 1984, ApJ, 286, 644

PACZYNSKI, B. 1997, ApJ, 11, 53

PERLMUTTER, S. ET AL 1997, ApJ, 483, 565

PERLMUTTER, S. ET AL 1997, astro-ph/9812133

PHILLIPS, M. M. 1993, ApJ, 413, L105

PINTO, P. A., & EASTMAN, R. G. 1996, astro-ph/9611195

REINECKE, M., HILLEBRANDT, W., & NIEMEYER, J. C. 1998, astro-ph/9812120

RIESS, A. G., et al1998, AJ, 116, 1009

sc Schmidt, B. P., et al1998, ApJ, 507, 46

TURNER, M. S., & WHITE, M. 1997, PRD 56, R4439

UMEDA, H., NOMOTO, K., YAMOAKA, H., & WAJANO, S. 1998, astro-ph/9806336

VAN DEN HEUVEL, E. P. J., BHATTARCHARYA, D., NOMOTO, K., & RAPPAPORT, S. 1992, A&A, 262, 97

YUNGELSON, L., & LIVIO, M. 1998, ApJ, 497, 168